AF482541

European comparative public-private partnership study (Germany)

A study of the economic efficiency of the life cycle costs of over 1 000 conventional buildings and 18 public-private partnership projects in Germany

Jörg Christen, Mainz

Status: 05 December 2024

About the author

Dr Jörg Christen has been working on the subject of public-private partnerships (PPPs) in various capacities. From 1991 to 1994, he was responsible for a series of pilot property leasing projects at the Rhineland-Palatinate Ministry of Finance. From 2004 to 2022, he headed the PPP department there. At the national level from 1997 to 1999, he was co-head of the federal-state working group on economic efficiency studies for parallel tenders (hire purchase, leasing and rental) at the Federal Ministry of Regional Planning, Building and Urban Development. After 2001, he became project coordinator of the PPP Steering Committee for public private partnership projects in the public building sector under the leader-ship of Parliamentary State Secretary Achim Großmann, MdB. From 2004 to 2009, he served as the head of the German Federal PPP Task Force at the Federal Ministry of Transport, Building and Urban Development. Since 2000 he is working as a lecturer at Mainz University of Applied Sciences and has supervised several master's and bachelor's theses on "Evaluating the economic viability of PPPs" at the Faculty of Engineering. Recently, he has also collaborated with other universities of applied sciences and universities (Münster University of Applied Sciences, Karlsruhe Institute of Technology/Technical University of Karlsruhe). Contact: joerg.christen@web.de.

*

Imprint

© 2024 Dr Jörg Christen, Mainz

Cover picture credits:

1 row:	Bernd Lohse, Dohle+Lohse	Wolfgang Kariger	Richard Stöhr
2 row:	Karl Müller	Logo EU	Johannes Seyerlein
3 row:	Michael Voigt	Logo Deutschland	Markus Lugert
4 row:	emptyform/tjie	Christoph Schroll	Christoph Schroll

Cover layout: Erik Kinting, Hamburg

Publishing and Printing: tredition GmbH, Ahrensburg

ISBN: 978-3-384-19672-9

Bibliographic information from the German National Library:

This publication is listed by the German National Library in the German National Bibliography; detailed bibliographic data are available online at http://dnb.d-nb.de.

PREFACE

When the Federal PPP Task Force was set up in the Federal Building Ministry on July 4, 2004 as a staff unit under the then Parliamentary State Secretary Achim Großmann, MoB, this was preceded by a three year start-up phase that had begun with the so-called Kanzler AG, from which the so-called PPP Steering Committee for Public Building Construction emerged, a body comprising federal and state ministries, municipal umbrella organizations, associations from the construction and credit industries, the Federal Chamber of Architects and trade unions. At the same time, an open working group chaired by Member of the Bundestag Dr. Michael Bürsch, MoB met in the Bundestag from 2003 to 2009. The leitmotif of this work was the objective of using the PPP lifecycle approach - i.e. the interlinking of planning, construction, financing and operation - to achieve efficiency benefits and to provide an impetus for administrative modernization by comparing the new procedure with the previous traditional procedure. The key to the start of the work was, on the one hand, experience from abroad (such as a very positive report from the British National Audit Office) and, on the other, a list of 46 German PPP precursor projects from the 1990s in which savings of 20% were reported. There was a great spirit of optimism in the work, people felt they were an active part of Agenda 2010.

The work of the PPP Task Force showed that the objective of increasing efficiency was taken seriously: First of all, two interministerial working groups of the federal and state governments meeting in parallel, with the participation of the audit offices, developed a measurement procedure with the guideline "Economic feasibility studies for PPP projects in public building construction" and, in cooperation with the PPP Task Force of the state of North Rhine-Westphalia and its head Dr. Frank Littwin, presented it to the Conference of Finance Ministers in 2006; the principles laid down here are to be applied to every PPP procedure. In addition, the website www.ppp.projektdaten¬bank.de was launched, in which the results of PPP projects are to be presented transparently. The federal PPP task force selected 10 pilot projects, which were provided with evaluation clauses to enable the economic viability of the projects to be reviewed during the operating phase. In addition, 2008 saw the launch of an evaluation program for 50 school projects as a first step towards ongoing evaluation.

This focus on economic efficiency soon bore its first fruits: in a DIFU survey in 2005, 83% of the 231 municipalities and 80% of the 63 rural districts surveyed cited the expectation of efficiency gains as the main reason for implementing PPP projects. The PPP project database reports the results of 118 of the 281 projects to date, with average savings of 13% (between 1 and 32%). However, there is a lack of transparency and credibility because the studies are not publicly accessible. This has also been a focus of criticism in recent years, which ultimately became a major reason for the drastic decline in the annual number of projects in building construction: while there were 32 projects in the peak year of 2007, there were only four contracts signed in 2019. In 2023, however, the number of projects rose again to ten, with the highest PPP investment costs per year measured to date (EUR 1.2 billion). It will be interesting to see whether this is just a catch-up effect of the pandemic or whether the very good results of the 2019 PPP School Study may have triggered a change in sentiment among decision-makers.

Against this background, I am now very pleased to be able to present the results of a cross-sectional study on the economic efficiency of the PPP life cycle approach of 18 PPP projects for the first time 20 years after the founding of the PPP Task Force of the Federal Government. I am very grateful that, following the transfer of the PPP Task Force to Partnerschaften Deutschland AG in 2009, I was able to carry out this evaluation work through my job at the Rhineland-Palatinate Ministry of Finance and my lectureship at the Chair of Prof Dr Ulrich Bogenstätter at Mainz University of Applied Sciences, Faculty of Engineering. The current and previous work - economic efficiency studies on individual PPP projects, daycare center study 2015, PPP school study 2019, PPP maintenance and operator liability 2020 (in cooperation with Münster University of Applied Sciences) and PPP and energy efficiency 2022 (in cooperation with TU Karlsruhe/KIT) - benefited greatly from the commitment and enthusiasm of the students in their Bachelor's and Master's theses, which I have been able to supervise in recent years.

In the 2019 PPP school study, the focus was placed on the connection between planning, construction and maintenance, because maintenance costs represent the largest block of operating costs after financing costs and the PPP life cycle approach with the interlinking of planning, construction and operation can be particularly well illustrated here. This study now extends the scope of the investigation to the entire life cycle costs over the agreed contract period. Not only 16 PPP school projects were examined, but for the first time also two projects from the administrative buildings sub-sector. The topic of energy efficiency in PPP projects was dealt with in parallel as part of a master's thesis written at the KIT Institute of the Technical University of Karlsruhe, the partial results of which have been incorporated into the present study.

Just as with the PPP school study (2019), the KGSt - Kommunale Gemeinschaftsstelle für Verwaltungsmanagement - provided anonymised data from 816 school buildings and 174 administrative buildings from the building management comparison analysis; we are very grateful to the KGSt for this, particularly as, at the European level, it has proven difficult to obtain data from conventional projects. To calculate the life cycle costs over 25 years, BKI key figures were also applied to the construction costs for the KGSt model. The results obtained from this data are referred to as the KGSt model. It should be stressed that the KGSt values used in the study and the so-called KGSt model were neither calculated nor interpreted by the KGSt.

This study would not have been possible without the support of a large number of people: We would therefore like to thank in particular the PPP companies GOLDBECK Public Partner GmbH, HOCHTIEF PPP Solutions GmbH, VINCI Facilities Solutions GmbH and ZECH Facility Management GmbH, who made their project contract documents available. We would also like to thank the employees of 13 local authorities who gave consent for their project to be included in the study and were on hand to answer any questions.

Special thanks are due to Professor Dr-Ing. Ulrich Bogenstätter for his support and expert advice, as well as to Professor Dr-Ing. Dipl.-Wi.-Ing. Kunibert Lennerts and Dr-Ing. Dipl.-Wi.-Ing. Heike Schmidt-Bäumler and Mr Sebastian Vöst for their cooperation in the creation of this master's thesis. The same applies to Professor Dr Frank Riemenschneider-Greif, Münster University of Applied Sciences, for the bachelor thesis on operator liability in PPP maintenance written by Mr Konstantin Scheidt under his supervision, which provided important information on the current situation in conventional and PPP maintenance management by surveying over 300 local authorities.

Finally, I would like to sincerely thank my colleagues at the European PPP Expertise Centre (EPEC), based at the European Investment Bank (EIB), and in particular to Dr Aris Pantelias, who oversaw the work on the European PPP comparative study.

Last but not least, I would like to thank my former, highly esteemed chief, Mr Achim Großmann, Parliamentary State Secretary in the Federal Ministry of Construction, MoB, as well as to the head of the PPP working group in the German Bundestag, Dr. Michael Bürsch, MoB, and my long-time colleague Dr. Frank Littwin, who unfortunately passed away far too early in recent years and who would certainly also have been very pleased with the good results of this study.

Mainz, December 2024

The author

PREFACE

This paper is the latest milestone in a series of studies on the economic viability of the PPP life cycle approach that have been conducted over the last ten years by Dr Jörg Christen in the Building and Property Management/Facilities Management course at the Faculty of Engineering at Mainz University of Applied Sciences in connection with master's and bachelor's theses. These include several studies on the economic viability of individual PPP projects, cross-sectional studies on the management of maintenance in daycare centres (PPP daycare centre study 2015) and schools (PPP school study 2019) or on specific individual topics such as operator liability for maintenance (bachelor's thesis with the Münster University of Applied Sciences, 2020) or PPP energy efficiency (master's thesis with the Technical University of Karlsruhe (Karlsruhe Institute of Technology), 2022). Parts of this preparatory work were also incorporated into this study. As a result, this study was carried out on the basis of contract and operating data from 18 PPP projects involving 41 buildings with a total gross floor area of over 400 000 square metres as well as data from over 1 000 conventional buildings.

This is an important empirical contribution to a central area of research in my department, which focuses on part-specific life cycle costs and developed the life cycle cost calculator NUKOSI — calculation and simulation of utilisation costs — for this purpose.

In light of this, we would like to thank the Kommunale Gemeinschaftsstelle für Verwaltungsmanagement (KGSt) for providing the operating data for 990 conventional school and administrative buildings as well as the companies GOLDBECK Public Partner GmbH, HOCHTIEF PPP Solutions GmbH, VINCI Facilities GmbH and ZECH Facility Management GmbH for providing PPP contract documents and operating reports. This is extremely valuable because it provides the basis for fact-based expert discussions on the allocation of life cycle costs. The analysis of the usage costs from private PPP companies is also particularly exciting because the PPP fees were calculated under competitive conditions and therefore must be sustainable over 25 years for the projects to be completed successfully.

I would also like to acknowledge and thank Dr Jörg Christen, who in this study was able to substantiate the hypothesis on the basis of impressive data showing that the incentive structures embedded in public-private partnerships when using the life cycle approach are the most important driving force for efficient construction and operation and, in some cases, significantly above-average quality.

In conclusion, I would be very pleased if the results of this paper generated momentum for a constructive expert discussion on the efficient management of life cycle costs, the importance of maintenance for useful life periods and residual values, sustainable energy management and also for further evaluations of the PPP life cycle approach.

Prof. Dr-Ing. Ulrich Bogenstätter
Mainz University of Applied Sciences, Programme Director
BUILDING AND PROPERTY MANAGEMENT/
FACILITIES MANAGEMENT

PREFACE*

The performance comparison between projects delivered conventionally and those delivered via the PPP model has been of interest to practitioners and academics alike for many years. This interest has increased in recent years, both because of the establishment of the PPP model as a 'standard' tool in the delivery toolbox of many public authorities in Europe and around the world but also because of the interest of various stakeholders in determining the efficiency and effectiveness of this delivery model compared to 'business as usual'.

During the last 15 years, numerous studies have emerged on this subject matter originating from public authorities, audit offices, professional bodies as well as academic institutions. The conclusions have been mixed and often inconclusive. Central to the lack of a decisive conclusion has usually been the lack of sufficient or sufficiently representative data to analyse and compare between PPP and non-PPP project performance, particularly when it comes to analysing and comparing this over the investment life cycle.

In light of all this, the European PPP Expertise Centre (EPEC) of the European Investment Bank (EIB) started in 2018 collecting date for a comparative performance analysis of PPP and conventionally delivered projects. In this context, EPEC provided the impetus for Dr Jörg Christen at Mainz University of Applied Sciences to conduct a comparative study of PPP and conventionally constructed schools and administrative buildings in Germany. EPEC followed the work done by Dr Christen and acted as a sounding board during the years of its data collection and analysis.

For the avoidance of any doubt, EPEC did not participate in the actual data analysis for data protection reasons nor audited the relevant calculations and as a result cannot advocate for the accuracy of the results reported in this study. However, having followed the data collection efforts and the analysis approach used, it considers it one of the most complete whole-of-life comparison between PPP and non-PPP projects done to-date.

While the reported results concern primarily the specific sub-sector of schools as well as initial administrative buildings, the study still points towards an approach that could helpfully inform the analysis of other sectors to compare PPP and non-PPP project performance and arrive at relevant conclusions.

We sincerely hope that this work will be considered seriously by practitioners and academics as an example of how such a comparative study can be approached and hope to see it replicated and/or adapted to other sectors in the future. The delivery of public works has always been an area marred by lack of performance data and spending inefficiency. Being able to approach this in a scientifically robust way remains largely the holy grail to assess and ultimately promote the efficient and effective use of public funds.

Dr. Aris Pantelias & Edward Farquharson
EPEC – European PPP Expertice Centre
European Investment Bank

*The views, interpretations and conclusions contained in this Preface reflect the current views of the author(s), which do not necessarily correspond to the views or policies of the EIB or any EPEC member

FORWORD

Dear readers,

If a local authority takes a strategic approach to property maintenance, it can achieve significant potential savings over the entire life cycle (Lennerts et al., 2006[1]).

For over 20 years, a number of research projects undertaken by the Chair of Facility Management at the Institute for Technology and Management in Construction (TMB) at the Karlsruhe Institute of Technology have successfully engaged with systematic data collection and analysis as well as optimising decisions over the property life cycle.

For example, since 2001 the OPIK research project has been working continuously on the analysis and optimisation of facilities management processes in 20 hospitals and an open benchmarking study is in progress. The transparent depiction of processes with a product-based approach facilitates comprehensive performance benchmarking and the identification of potential for optimisation in energy and resource management as well as in-depth discussions on particular topics such as strategic alignment and key figures for maintenance staff.

The objectives of the BEWIS and EKiBA research projects, among others, are the analysis and optimisation of the management strategies of larger property portfolios, focusing on the various ongoing maintenance and repair measures for buildings and their impact on value retention. Specifically, the goal is to create a robust data basis to develop sustainable and efficient management strategies. In addition to reducing costs and extending useful life, it is also possible to sustainably reduce resource consumption and minimise life cycle costs through the targeted planning of interventions.

The results achieved by these research projects, which are just a couple of examples, once again demonstrate the enormous importance of a sound data basis.

The uncertainty created by the lack of comparative data at the German national and European levels has also led to repeated debates on the economic viability of PPP projects. In this context, the primary objective of the study, led by Dr Jörg Christen, is to establish a sound data basis for comparing PPP projects with conventional projects.

 The results of this extensive study are remarkable:

- Economic advantages: The analysis found that 18 PPP projects could offer potential life cycle cost savings of up to around €330 million over a 25-year period compared to conventional approaches.

- Efficient construction processes: PPP projects feature extremely efficient construction processes that reduce construction costs by 15-20%, reduce construction times by 30% and provide unrivalled certainty in terms of costs and deadlines.

- Superior quality: PPP projects have considerable advantages in terms of quality, particularly with regard to property organisation, maintenance and energy management.

- Promising results: The study indicates that PPP new-build projects are able to achieve better energy efficiency than conventional new builds.

 These results emphasise the fact that the PPP life cycle model has unlocked significant potential for efficiency in the PPP projects studied. This study supports the conclusion that appropriate incentive and liability mechanisms in PPP contracts can have an overall positive effect. The results of the study therefore suggest that greater emphasis should be placed on incentive structures, for example, for reducing energy consumption, which is something that was developed in detail in the master's thesis on PPP energy efficiency (in cooperation with Mainz University of Applied Sciences).

In summary, this study shows that PPP projects have considerable potential with regard to energy efficiency and cost optimisation. We hope that these results will inform and enhance the debate on

---

[1] Lennerts, Pfründer and Bahr (2006). Lebenszyklusorientierte ganzheitliche Unterhalts- und Instandhaltungsstrategien für Schulen.

PPP projects and their economic viability. We would like to emphasise that this paper may spark the sustainable development of databases on the construction and operation of PPP projects — in cooperation with the Karlsruhe Institute of Technology — as has already been done several times.

Univ.-Prof. Dr-Ing. KUNIBERT LENNERTS
Karlsruhe Institute of Technology (KIT)
Institute for Technology and Management in Construction

Contents   I

List of tables

List of abbreviations

II. BV	Second German Calculation Ordinance
AMEV	Working group for mechanical and electrical engineering in state and local authorities (Arbeitskreis Maschinen- und Elektrotechnik staatlicher und kommunaler Verwaltungen)
BBSR	German Federal Institute for Research on Construction, Urban Affairs and Spatial Development (Bundesinstitut für Bau-, Stadt- und Raumforschung)
GFA	Gross floor area
CC	Construction costs (DIN 276 excluding CG 100 and CG 760)
BKI	Building cost information centre of the German Chamber of Architects
BNB	Evaluation system for sustainable construction (Bewertungssystem Nachhaltiges Bauen)
BWI Bau	Institute of the Construction Industry (Institut der Bauwirtschaft)
DAB	German architect's journal (Deutsches Architektenblatt)
DESTATIS	German Federal Statistical Office (Deutsches Statistisches Bundesamt)
DIN	German Institute for Standardisation (Deutsches Institut für Normung)
DIN 276	Cost planning for building construction; CG 100 Property; CG 200: Preparation and development; CG 300: Building construction; CG 400: Technical building facilities; CG 500: Outside facilities; CG 600: Furnishings and artworks; CG 700 Additional building costs; CG 760 Financing costs
DIN 18960	Usage costs for building construction; CG 100: Capital costs; CG 200: Property management costs; CG 300: Operating costs; CG 350: Operation, inspection, maintenance; CG 400: Repair costs
DIW	German Institute for Economic Research (Deutsches Institut für Wirtschaftsforschung)
EPEC	European PPP Expertise Centre
ESA	European System of Accounts
Euribor	Euro Interbank Offered Rate
FMK	Conference of finance ministers of the German federal states (Finanzministerkonferenz der Länder)
FMK Guidelines	Guideline "Wirtschaftlichkeitsuntersuchungen bei PPP-Projekten" (2006) ("Economic viability studies for PPP projects")
GEFMA	German Facility Management Association
GWB	German Act against Restrictions on Competition (Gesetz gegen Wettbewerbsbeschränkungen)
ISFR	International Financial Reporting Standards
CPP	Conventional procurement process
CG	Cost group
KGSt	The company Kommunale Gemeinschaftsstelle für Verwaltungsmanagement
KGSt actual	Maintenance budget for the KGSt benchmark perimeter buildings
KGSt target	KGSt target figures for calculating the maintenance budget
MinBl	German Ministerial Gazette (Ministerialblatt)
NVwZ	Neue Zeitschrift für Verwaltungsrecht journal
P2P	Public-public partnerships
OSCAR	Office Service Charge Analysis Report
PABM	Practice-oriented adaptive budgeting of maintenance measures Bahr/Lennerts, Technical University of Karlsruhe
PPP	Public-private partnership
RBBau	Guidelines for the execution of German federal construction (Richtlinien für die Durchführung von Bauaufgaben des Bundes)
RLBau	Guidelines for the execution of German regional construction (Richtlinien für die Durchführung von Bauaufgaben des Landes)
VDI	Verein Deutscher Ingenieure (Association of German engineers)

Summary

(1) The present study of the life cycle costs of 18 PPP projects shows clear economic advantages and a high level of user satisfaction for the PPP projects after approx. 55 per cent of the contract term in comparison with BKI key figures and KGST operating data for 814 school and 176 administrative buildings:

European Comparative PPP Study - Germany			GREENFIELD PROJECTS Projects 1-16		BROWNFIELD PROJECTS Projekte 17-18		ALL PROJECTS Projekte 1-18	
			PPP/KGST	PPP/BKI	PPP/KGST	PPP/BKI	PPP/KGST	PPP/BKI
DIN 276	CG 200-700	Construction costs	-17% (-15% / - 20%)		-11%		-16% (-13% / - 19%)	
		Cost overruns (n=15)	0,6% (Median 0%)		3,5%		0,7% (Median 0%)	
DIN 277		Construction time	-30%		-29%		-30%	
		Time overruns (n=15)	2,3% (Median 0%)		0%		2% (Median 0%)	
DIN 18960	CG 100	Capital costs	-15%	-15%	-8%	-8%	-14%	-14%
	CG 200	Object management costs	12%	23%	18%	27%	14%	24%
	CG 300	Operating costs	19%	-24%	33%	-5%	20%	-20%
	CG 310	Water, heating, electricity	-16%	-33%	11%	-10%	-15%	-30%
	CG 320	Disposal costs	0%	0%	0%	0%	0%	0%
	CG 330/40	Cleaning costs	5%	9%	-15%	-14%	2%	6%
	CG 350	Servicing&inspection	685%	-21%	0%	0%	685%	-21%
	CG 360	Energy management	0%	-2%	0%	0%	0%	-2%
	CG 370	Taxes, contributions, assurance	179%	-7%	211%	-22%	179%	-11%
	CG 390	Other operating costs*	455%	1660%	7310%	5955%	455%	1660%
	CG 400	Repair costs	66%	-10%	69%	-21%	68%	-9%
	CG 200/350/400	Maintenance budget	137%	-15%	80%	-40%	128%	-17%
		in % p.a. PPP / KGST-TARGET	1,6%	1,2%	1,8%	1,2%	1,6%	1,2%
		in % p.a. KGST-ACT / BKI	0,6%	1,7%	0,6%	1,7%	0,6%	1,7%
Usage costs without risk costs			-1%	-15%	11%	-7%	0%	-14%
Usage costs including risk costs			-3%	-15%	10%	-7%	-2%	-13%
Residual value			27%	0%	30%	0%	27%	0%
Life cycle costs without risk costs			-3%	-32%	39%	-17%	2%	-30%
Life cycle costs including risk costs			-35%	-34%	-17%	-20%	-32%	-32%
Life cycle costs including risk costs and VAT			-37%	-36%	-21%	-23%	-34%	-34%

* Percentage of the cost group CG 390 in the usage costs: PPP 3.2% (of which approx. 3% costs are attributable to KG 200), KGST: 0.3%, BKI: 0.2%

Table 4.1: Comparison of life cycle costs PPP vs. KGST/BKI -new-build and renovation projects

1.　PPP new-build projects: 14 school and 2 administrative building projects

(2) The life cycle costs of the PPP new-build projects show an advantage of 35% and 34% over the KGST and BKI variants respectively, often with above-average qualities and taking into account the expected building residual values over 25 years.

PPP new-build benefit: 34%/35% (KGSt/BKI): Life cycle costs including residual values

(3) The pure usage costs, without consideration of qualities and risks, are between 1 % and 15 % below the costs of the KGST and BKI variants. The result for the life cycle costs including residual values is significantly better than the pure cost comparison, as the different maintenance strategies with their effects on structural damage, useful life and residual values are taken into account.

1%-15% for pure usage costs without consideration of risk and quality

(4) The PPP construction costs indicate average savings of 17% (15-20%) over the conventional comparable BKI values with average to above-average quality levels, combined with remarkable cost certainty. Additionally, construction times are 30% shorter with a high degree of deadline certainty. These savings resulting from the efficiency of the construction process allow, in particular, the implementation of a high-quality maintenance strategy.

Construction costs: - 17% (15% to -20%) < BKI Construction time: 30% < BKI

(5) PPP maintenance management has significant qualitative advantages over the reality of conventional procurement. PPP projects have budgets sufficient for a medium to above-average level of maintenance during the agreed operating phase, determined based on part-specific maintenance calculations, as well as service levels specifying the required qualities of the key elements, user requirements with fixed response and correction times and fee reductions if target specifications are not achieved, as well as a reserve account for rapid access to funds, which has been described as a "quantum leap" by conventional construction management staff.

Efficient construction processes and contractual incentive structures enable high-quality maintenance

(6) Energy management is characterised by contractual rules involving the assumption of risk for guaranteed maximum consumption volumes by the PPP company as well as participation in savings. This results in savings of 11% which have actually been achieved or can be forecast over the remaining term of the project in comparison to the costs calculated based on the maximum consumption values. Heat consumption is 38% below the VDI 3807 guideline value, 58% below the VDI 3087 average value and 56% below the comparable KGSt values; the guaranteed maximum consumption values were calculated quite ambitiously. Incidentally, the savings in thermal energy consumption increase in proportion to the PPP company's share in the savings.

Good energy management

Incentive structure: Participation in savings by the PPP company shows positive effect

(7) PPP projects clearly prioritise property management budgets; unlike the KGSt model (no. 5) or BKI model (no. 6), these are the third largest cost item for public-private partnerships. This is explained by the incentive and liability structures intrinsic to public-private partnerships. On the other hand, it should be noted that the PPP additional costs in this cost group (12% compared to KGSt or 23% compared to BKI plus a part of the other operating costs) also include VAT, which is not payable under conventional approaches and results in additional income for the federal, state and local authorities in the case of public-private partnerships.

PPP focus on property management

(8) PPP capital costs are on average 15% lower due to lower construction costs. However, interim financing and final financing have higher interest rates (forfeiting with waiver of defence: +0.18%; project financing: +0.59%). On the other hand, there are qualitative advantages such as guaranteeing the fixed construction cost price through interest rate hedging for interim financing and additional quality assurance by the refinancing bank for cases with project financing.

Capital costs Higher PPP interest rate Consideration for risk hedging

(9) PPP cleaning costs are higher than the KGSt (+5%) and BKI (+9%) estimates. This partial result may change, however, if cleaning intervals are factored in. The analysis of a PPP project revealed that the catalogue of services was 30% more extensive than DIN 77400 and that the PPP disadvantage in cleaning costs per m^2 of gross floor area was translated into a PPP advantage per m^2 of annual cleaning area compared to the KGSt key figures. This type of qualitative review could not be carried out for the other projects in the timeframe for this study.

PPP cleaning costs per m^2 gross floor area 5%-9% higher

(10) The considerable percentage increase in PPP costs under other operating costs includes cost items from other cost groups and refers to relatively low conventional nominal costs, which account for only 0.3% (KGSt) or 0.2% (BKI) of conventional usage costs.

Other operating costs; large percentage differences, low share of total costs

(11) In addition to costs, economic viability studies must also assess project risks. On the basis of the results of the PPP school study (2019), the first data quantifying the maintenance risk are now available, making it possible to quantify the risk of structural damage due to neglected maintenance (at 2% of the restoration costs) and the effects of different maintenance budgets on the future residual value. The PPP advantage over the KGSt model is then increased to 3% for usage costs. Since the KGSt maintenance budgets are 50% lower than the target values for a moderate maintenance level, this results in shorter useful lives and correspondingly lower residual values. Conversely, useful life and residual value increase when maintenance budgets are above average. On this basis, the expected residual value of the PPP projects is on average 27% higher than the KGSt alternative and on average the same as the BKI alternative.

Risk assessment: Maintenance

Too low budgets: Structural damage and reduced residual value

Higher budgets extend use and increase residual value

Residual value: PPP/KGSt: +27% PPP/BKI: -1%

(12) When comparing the life cycle costs (usage costs minus residual value) without a risk assessment, there is a PPP advantage of 3% over the KGSt model and 32% over the BKI model. When the risk assessment is included, the PPP advantage increases to 35% over the KGSt model and to 34% over the BKI model.

Life cycle costs excluding/including risk PPP/KGSt: -3%/-35% PPP/BKI: -32%/-34%

2. Renovation projects: 2 school projects

Renovation projects

(13) For both renovation projects, the life cycle costs show an advantage of 17% and 20% over the KGST and BKI variants respectively. In terms of pure usage and life cycle costs without taking qualities and risks into account, the results of the PPP projects are 11% and 39% higher than the KGST variant and 7% and 17% lower than the BKI variant.

Life cycle costs: PPP/KGST: -17% PPP/BKI: -20%

(14) However, in this case the PPP construction cost advantage is only 11%; it should be noted, that a detailed analysis of the renovation work could not be carried out within the scope of this study. This also applies in particular to the assessment of the risk of additional work in historical conservation projects: In one PPP historical conservation renovation project, the subsequent cost increases amounted to only 3.5%, which represents excellent construction performance compared to similar projects implemented using conventional approaches.

Problem: Comparison of construction work

(15) In addition, the two PPP renovation projects have higher property management and supply costs than the new-build projects, but lower cleaning costs.

3. Overall result

(16) The life cycle costs of all 18 PPP projects show an advantage of 17% to 35% (on average 32%) over the KGST and BKI variants, often with above-average qualities and taking into account the expected building residual values over 25 years.

Life cycle costs incl. residual values PPP vs. KGST/BKI: -17% to -35% (Ø 32%)

(17) The very efficient construction process enables maintenance budgets to be increased by more than 50% compared to the KGST variant, while usage costs are the same (PPP/KGSt) or 14% (PPP/BKI) lower.

Pure usage costs: PPP/KGSt: 0% PPP/BKI: -14%

(18) The participating local authorities rated the PPP performance of construction and operation at 1.7. (1= very good, 5 = unsatisfactory).

Local authorities rate PPP construction and operation at 1.7

(19) There are some clear qualitative advantages in maintenance, energy management and property organisation.

High-quality maintenance+energy+property management

(20) The high-quality PPP maintenance management leads to significantly higher residual values at the end of the contract compared to the KGST variant.

Higher residual values PPP/KGST: +27%

(21) The result for public-private partnerships is also improved by 2% if the additional VAT income for the federal, state and local authorities that arises from the VAT charges on PPP personnel costs is taken into account.

Additional VAT income for PPP approx. 2% of the life cycle costs

(22) It should be noted that indexing has a not inconsiderable impact on the overall result. This applies especially to the construction price index, which is used to index maintenance costs and future residual values. If the construction price increase rate of 2.92% per annum calculated for the period from 2007 to 2021 were instead 2%, the PPP advantage in terms of life cycle costs including risk assessments would fall from 35% to 26% compared to the KGSt model and from 34% to 28% compared to the BKI model. With an annual increase in construction prices of 1.5%, the PPP advantage is reduced to 23% (KGSt) and 27% (BKI). In this respect, rising inflation increases the significance of the maintenance risk and its impact on structural damage and changes in residual value.

Effects of indexing

Rising inflation increases residual value and PPP advantage

(23) One of the key reasons for these positive results lies in the PPP contractual incentive and liability structures[1]. The PPP company is in a long-term contractual relationship, bears the price and schedule risks for construction costs and maintenance and guarantees the maximum energy consumption volumes. The PPP company carries the operator risk and is therefore motivated to optimise organisational and operational structures and minimise project risks so as to safeguard its opportunities to make a profit. This is working with the life cycle approach. There are a whole range of organisational differences to the conventional procedure in this regard.

Reason for PPP advantages: Incentive and liability system in the PPP contract

4. Outlook

(24) The good PPP results of the present studies not only justify the legitimisation of the launch of new projects, they also show how important evaluation work is. They enable a fact-based assessment of the various alternative courses of action, highlight their strengths and weaknesses and thus provide an impetus for improving the status quo.

Good results legitimise new projects

Evaluation provides impetus for modernisation

(25) This applies, for example, to PPP maintenance management, the qualities of which were already determined as part of the PPP school study (2019) and confirmed in this study: In the PPP projects analysed, there will in all likelihood be

PPP efficient tool against investment backlog

no maintenance backlog at the end of the contract term. PPPs are therefore proving to be an efficient tool for reducing the municipal investment backlog, which according to the KFW Municipal Panel will have grown to €186 billion by 2024.

(26) As not all projects can be handled via PPP in the future, it will be important in this context to introduce structural improvements in conventional maintenance management. This study contains a number of suggestions for this (e.g. introduction of component-specific maintenance calculations in the design planning, definition of service levels for the main components with response and repair times; establishment of reserve accounts for rapid drawdown of funds; conversion of financing from full to partial amortisation in accordance with the so-called so-called golden balance rule; greater attention to the correlation between maintenance budget and residual value; comparison of indexed cost curves and residual values in profitability analyses in addition to present value analyses).

Suggestions for improving conventional maintenance

(27) A key finding of this study is the connection between contractual incentive structures and good PPP performance. PPP is thus in the research field of the 2016 Nobel Prize in Economics, which was awarded to Professors Bengt Holmström and Oliver Hart for their work on principal agent theory and incentive mechanisms in long-term contracts. It remains to be seen whether it will be possible to anchor similar incentive mechanisms in the conventional process (e.g. performance-related fees; user claims for compliance with response and operating times that are subject to sanctions or, for example, a participation of administrative units in energy savings).

Contractual incentive structures are efficiency drivers

Introduction also in the conventional procedure ?

(28) An alternative to this is to introduce legal requirements for the publication of key figures on the estimated maintenance budget and the provision of minimum maintenance budgets.

Alternative: legal anchoring

(29) In any case, it would be desirable for the present study to also provide an impetus for placing the research and evaluation work and the transfer of knowledge on the PPP life cycle approach on a more sustainable organisational structure in future.

Put evaluation and knowledge transfer on a sustainable organisation structure

1 Introduction
1.1 Background

(1) The goal of public-private partnerships is the sustainable optimisation of costs, quality and processes over the life cycle of public infrastructure projects[2]. Responsibility for the planning, construction, financing, operation and, where necessary, use of a school, nursery, sports hall or section of road is handled by an interdisciplinary private PPP contractor for a long-term contract period. The goal is to generate cost savings with a high degree of cost certainty, shorter construction times with high deadline certainty, improved management of maintenance with sustainable quality control for the duration of the contract, improved operating costs (for example, reduced energy consumption and cleaning costs) by linking construction and operation throughout the planning process, and improvements in revenue and residual value compared to the conventional status quo.

Objectives Increased efficiency from life cycle approach

(2) However, in recent years PPP projects have been strongly criticised: "Everything was supposed to be better, almost none of it happened" (*Süddeutsche Zeitung*, 13 June 2014); "Private construction more expensive" (*Handelsblatt*, 12 June 2014); "The sold state" (*Die Welt*, 9 February 2014); "The plundered state — secret deals between politics and business", television programme on Arte, 11 February 2014). Significantly higher private financing costs[3] and the requisite additional profit and risk costs made public-private partnerships unprofitable compared to conventional projects. Economic viability studies that state otherwise allegedly lack transparency[4] and are calculated incorrectly[5]. In reality, the main driver of public-private partnerships is not increased efficiency, but budget modelling, because public-private partnerships deliberately bypass the debt brake[6].

Criticisms of PPP

- *uneconomical*
- *financing too expensive*
- *risk premiums too high*
- *non-transparent*
- *bypassing the debt brake*

(3) Accusations of inefficiency are in contradiction to the findings of published PPP economic analyses: For 118 of the 281[7] projects in the building construction sector to date across Germany with an investment volume of over €9.6 billion, the findings of the economic viability studies carried out before contract conclusion were published in the German government's PPP project database. According to these findings, in these projects efficiency increased by an average of 13% compared to the conventional status quo, with individual values ranging between 1% and 32%. However, the economic viability studies that lead to the results are only discussed within the respective administrative bodies and are generally not published. As a result, an important element required for public traceability, transparency and research credibility is missing.

Interim status of 118 projects: Ø 13% (1% - >32%) Life cycle cost savings over 25 years

(4) Lastly, a central point of criticism has arisen from a joint experience report by the German federal and state audit offices on the economic viability of PPP projects published in 2011 with a collection of audit reports on 17 PPP projects[8]. Four of the projects had a positive audit result confirmed and four projects had a positive audit result corrected to a negative result. For seven projects — i.e. the majority of the projects — it was found that the calculated economic benefit of public-private partnerships was not proven conclusively. Specifically, there is insufficient data.

Criticism from the Audit Offices: Lack of reliable data, PPP advantages implausible

(5) Following years of debate with the State Audit Office on the first local authority PPP projects, the state parliament of Rhineland-Palatinate passed a resolution in 2014 to compile a database of conventional and PPP projects sufficient to provide a reliable basis for PPP economic viability studies.

Rhineland-Palatinate state parliament 2014: Creation of a comprehensive data-base of conventional and PPP projects

(6) In response, the PPP department in the Rhineland-Palatinate Ministry of Finance, in cooperation with Mainz University of Applied Sciences, Faculty of Engineering, and supported by operational data from the KGSt comparative study, initiated a series of evaluations of the economic viability of public-private partnerships. This included, in particular, studies on the economic viability of maintenance budgets for 370 conventional and seven PPP childcare centres, the economic viability study on the construction and maintenance of 880 conventional schools and 50 PPP school projects (PPP school study 2019) and an initial economic viability study on PPP life cycle costs using the example of the PPP vocational college in Duisburg.

PPP evaluation studies at Mainz University of Applied Sciences, Faculty of Engineering

1.2 Setting goals for the European PPP comparative study

(7) At the European level, the annual conference of the PPP organisations of the EU Member States in 2018 expressed concerns about the lack of a valid data basis for comparing PPP projects with projects carried out under conventional arrangements. In many countries, the resulting uncertainty has led to the repeated generalised accusation that PPP projects are fundamentally uneconomical, both in public debate and by audit offices. It was therefore decided that the European PPP Expertise Centre (EPEC), based at the European Investment Bank (EIB), would carry out a European PPP comparative study in cooperation with the Member States in order to create a robust data basis to enable comparisons to be made.

European comparative PPP study EPEC (EIB)

(8) The focus here will be on cost and deadline certainty and comparing the life cycle costs of conventional and PPP projects. Because of the preliminary work on the PPP school study, in Germany the work was carried out together with the Federal Ministry of Finance in cooperation with Mainz University of Applied Sciences and the specialist PPP department in the Rhineland-Palatinate Ministry of Finance.

Cooperation between Mainz University of Applied Sciences, Faculty of Engineering, and the Ministry of Finance of Rhineland-Palatinate

1.3 Approach

(9) This study of German PPP projects draws on contract documents, tender documents and operating reports made available for evaluation by the participating PPP contractors under confidentiality agreements, as well as site visits to four projects. There was also a survey on cost and deadline certainty in the construction and operating phases. The local authorities party to PPP contracts were asked for their consent for the research results to be included in the EU comparative study in an anonymised form. The work status of the individual projects, as summarised in the project documentation, was communicated to the local authorities in advance. The preliminary work for the study on the German projects began in the fourth quarter of 2020, with provision and analysis of the data taking place from July 2021 to August 2022. In 2023, the local authorities were also surveyed to assess construction and operating performance.

1.4 Dataset

(10) The PPP dataset was compiled from 18 PPP projects, with four PPP companies and 13 local authorities acting as contractual partners. These projects include 13 new-build projects, three combination new-build and renovation projects and two renovation projects. 16 projects (14 new builds and two renovations) are school projects with 27 buildings and 12 sports halls, in addition to two projects involving administrative buildings. Six projects involve several buildings and locations in one project contract, which means that the data from 41 buildings at 25 locations with a total gross floor area of 401 618 square metres was analysed. Two projects contained separate sections in the contract for different locations, allowing for separate analyses to be carried out.

PPP dataset: 18 PPP projects, 25 locations, 41 buildings

(11) The analysed PPP contracts were concluded between 2003 and 2018. The contracts have an average term of 24.6 years and the average percentage of the operating period already completed is 55%.

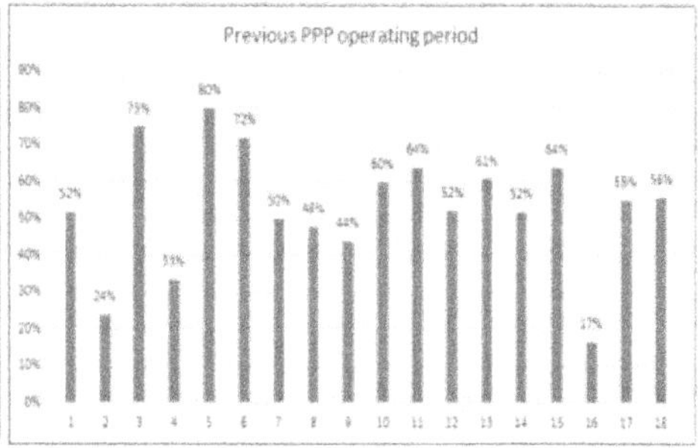

PPP-NEW+REN operating period	PPP
Minimum	17%
1st quartile	49%
Median (MED)	54%
3rd quartile	63%
Maximum	80%
Average value (AV)	53%
Average weighted value (AWV)	59%
Ø MED/AV/AWV	55%

n=16

PPP contract duration Previously Ø 49%

Table 1.1: Previous PPP operating periods

(12) PPP project contracts are primarily what is known as the PPP ownership model, in which the public client retains ownership of the land and property. In one project, the PPP company first acquired the land, constructed the building on it and leased the property, including the land, to the local authority. At the end of the contract term, the contract includes an option for the local authority to purchase the property. In another project, a leasehold contract with a long-term rental agreement and a reversion provision was agreed.

PPP contract model

(13) The prices of the PPP construction costs are fixed, including interim financing, and there are penalties for failure to meet the completion dates. For 16 projects, the financing of construction costs is part of the contract. Forfeiting with waiver of defence is planned in 11 cases and project financing in five cases. For two projects, the local authority is responsible for the final financing of the construction costs.

PPP financing structure

(14) The typical elements of a PPP contract include a long-term maintenance obligation based on service levels with a reserve account, energy management with guaranteed maximum consumption volumes for water, heating and electricity, cleaning services and property management. Failure to comply with performance obligations generally results in deductions from remuneration.

Content of PPP contract

(15) For the conventional procurement processes, the key cost figures from the BKI building cost information system of the German Chambers of Architects and the key cost figures from the KGSt — Kommunale Gemeinschaftsstelle für Verwaltungsmanagement — were used. This included data from 990 buildings (816 conventional schools projects, including 489 general education schools with a sports hall, 214 general education schools without a sports hall, 73 vocational schools with a sports hall, 26 vocational schools without a sports hall, 12 schools for children with special needs and 176 administrative buildings).

BKI figures and KGSt operating data from 990 schools and administrative buildings

1.5 Methodology

(16) In principle, the methodology of the study is based on the guideline "Wirtschaftlichkeitsuntersuchungen bei PPP-Projekten" ("Economic viability studies for PPP projects") (2006). To compare the life cycle costs of conventional and PPP projects, construction costs according to DIN 276, usage costs according to DIN 18960, procedures, qualities, risk transfer and value development of the property must be analysed.

FMK Guidelines 2006 Benchmark: Procurement (average values)

(17) According to the FMK Guidelines, the relevant benchmark for public-private partnerships is the reality of conventional procurement. In practice, conventional average values are used for this. In contrast, PPP projects should be compared with optimised self-builds, according to common opinion from audit offices. Under the assumption that the key figures provided by the Chambers of Architects represent recommendations for optimised solutions, whereas the KGSt comparison data reflect the constraints of the reality of local authority procurement, the PPP data are compared with two conventional procurement methods:

Requirements of audit offices PPP comparison with optimised self-builds

(18)   1) KGSt model with operating data from the KGSt comparative study, combined with construction costs and construction times according to BKI key figures

PPP comparison with two conventional procurement process models

(19)   2) BKI model with BKI key figures for construction costs, construction time and usage costs.

(20) For PPP projects, the annual costs incurred over the agreed contract duration are determined using the data in the contract.

(21) For conventional projects, there is no operating data available for individual projects over longer usage periods (e.g. 25 years). The comparable KGSt and BKI data represent the average operating costs of a number of comparable projects per year. For example, the comparable KGSt data comes from a survey of operating costs from 2014. This means that the key value for repair costs, for

Key BKI and KGSt figures

example, is the average value of, for instance, 800 schools from this survey year. The individual projects were constructed in different years, and the repair costs of the individual projects vary accordingly, in the same way that the maintenance costs of an individual project are never linear, but vary with peaks within individual years. This effect is reflected in the large case base. This increases the validity of using average value as a basis for a long-term comparison period. When comparing the costs of PPP and KGSt/BKI, therefore, the fundamental assumption is that the repair costs for 2014, for example, will be incurred at the level of the key figures over the entire comparison period of 25 years.

(22) The conventional key figures must be indexed for the same period as the PPP costs. If the period of the PPP contract extends from 2010 to 2035, the KGSt key value for 2014 is first indexed back to 2010. The individual cost groups are then indexed over 25 years. For this purpose, the individual cost groups' specific index values were applied consistently for the public-private partnership and conventional procurement process costs[9]. Sensitivity analyses were conducted to show the effects of price increases at different rates.

Indexation

(23) This allows the indexed total costs of the individual cost groups and the sum of all indexed costs over 25 years for the individual PPP, KGSt and BKI models to be determined and compared with each other. The overviews of the results for the cost comparison, the comparison of the qualities and the individual results of the 18 projects can be found in Appendix A1, A2 and A4.

Comparison of indexed total costs over the duration of the contract

(24) Where possible, the arithmetic average, the median and the mean weighted by gross floor area plus the mean of the aforementioned averages were used for the analysis. The differences in the overall result are detailed in Annex A2.

Statement of different average values

(25) This study does not engage in an additional cash value analysis of the total costs. Each project's timeframe is different, so the interest rates for the cost of capital vary considerably. According to the principle of the FMK Guidelines which states that the relevant discount rate should correspond to the long-term refinancing interest rate for the conventional model, there are therefore considerable differences in the cash values of the individual projects. Furthermore, it should be noted that the calculation of the cash value minimises the significance of future costs compared to previously incurred costs. This is particularly concerning given the impact of (overly low) maintenance budgets on the useful life and residual value of properties.

No cash value analysis

1.6 Risk assessment

(26) When analysing the economic viability of various alternative options, it is essential to account for the different levels of risk transfer. However, a sufficient data basis is required to quantify the corresponding risk costs.

(27) In this study, the results of the PPP school study (2019) were used to assess maintenance risks. The study used data from 880 conventional school projects and 50 PPP school projects to examine the risks and rewards of different maintenance budgets and organisational structures. This revealed that the KGSt generalised target recommendations for a moderate level of maintenance correspond to the analyses of part-specific maintenance calculations according to the BNB guidelines (evaluation system for sustainable construction). Based on reports from practical cases, the following can be concluded:

Quantification of maintenance risks

(28) (1) Lower maintenance budgets create a considerable risk of structural damage, shorter useful life and reduced residual values.

(29) (2) The KGSt figure for annual maintenance budgets as a percentage of restoration costs allows for mid-range maintenance levels.

(30) (3) Higher maintenance budgets have a positive effect on useful life and residual values.

(31) There was no particular valuation of financing risks from interim financing and the final financing of construction costs (especially for project financing projects). PPP construction costs, including the costs of interim financing, have a fixed price. As a result, the PPP company hedges the interest rate for the interim financing. However, if the cash flow in the construction phase is financed conventionally using cash loans, interest rate hedging does not take place. In this respect, the use of PPP interest rates, which are 60 basis points higher, results in a cost advantage for the conventional models, which is offset by the qualitative advantage of public-private partnerships. This difference in quality was not taken into account when comparing costs.

No estimate of financing risks

1.7 Scope and limitations of this study

(32) The following assumptions were made in the calculation:

- In the case of renovation projects and projects involving multiple sites, both public-private partnerships and the two conventional models were based on the assumption that use would begin once the entire project was completed.

Simplifying assumptions

- In the conventional model, it was assumed that the construction costs would be covered by interim financing and then, once use began, externally financed alongside the construction costs.

- For the conventional financing interest rates, the actual data of the conventional financing practices were not requested. Instead, these were derived from the mark-ups on the reference interest rate disclosed in the PPP price sheets and empirical values.

- The PPP-specific indexing rules were not accounted for (PPP price indexing often only applies every five years or after a certain threshold value has been exceeded). Standardised indexing values were used for public-private partnership and conventional procurement process models, which may deviate from the indexes agreed in the PPP contract.

- Public-private partnership and conventional procurement process waste disposal costs (DIN 18960 CG 320) and cafeteria costs were generally treated equally.

(33) Within the scope of this study, it was only possible to examine the qualitative aspects of individual cost types to a limited extent in certain cases:

Partially limited quality inspection

- A comparison of PPP construction work with key BKI figures was restricted in the case of renovation projects.

- When evaluating maintenance services (DIN 18960 CG 200/350/400), the results from the PPP school study (2019) were used for the school projects.

- For the utilities costs (DIN 18960 CG 320), it was only possible to analyse guaranteed and actual consumption volumes and costs for some of the projects.

- With regard to cleaning services (DIN 18960 CG 330/40), quality analyses could only be carried out for one project.

2   Results of the PPP new-build projects

2.1 Construction costs

2.1.1 Costs

PPP-NEW operating period	PPP/BKI	GFA
Minimum	-32%	80.421
1st quartile	-21%	11.051
Median (MED)	-15%	15.770
3rd quartile	-11%	30.720
Maximum	-4%	4.298
Average value (AV)	-16%	23.190
Average weighted value (AWV)	-20%	
Ø MED/AV/AWV	-17%	

n=16

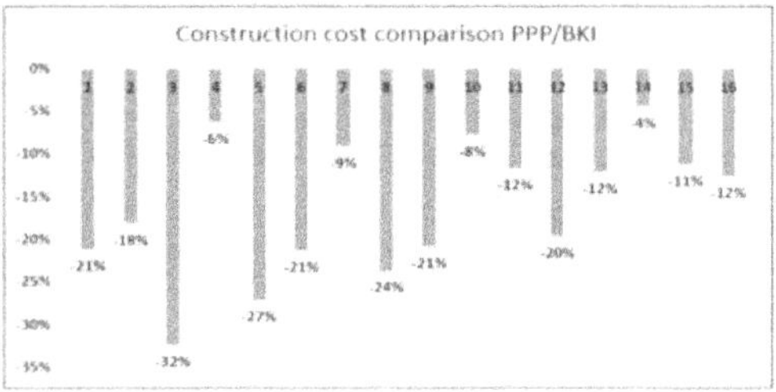

Table 2.1: PPP/BKI construction costs

a) RESULTS:

(34) In the 16 projects, the PPP construction costs are between 15% (median) and 20% (mean value weighted by gross floor area) below the BKI key figures. The advantage increases as the project volumes increase. Taking the average of the median, arithmetic mean and weighted mean value results in an average advantage of 17%. All the projects analysed are below the comparable BKI value, ranging from -4% to -32%.

PPP construction costs 15-20% under BKI

b) NOTES:

(35) The dataset of the case study group examined here consists of a total of 16 projects with a total gross floor area of 371 000 square metres. The projects include 14 school projects involving 25 buildings and 11 sports halls as well as two projects involving administrative buildings. The 16 projects comprise 13 new-build projects and three combination new-build and renovation projects.

(36) The comparable BKI costs include the DIN 276 CG 200-700 cost groups. The cost of land was not considered in this study. The corresponding comparable BKI values were used for the different building types (school building, sports hall, administrative building, new building/modernisation). Ancillary construction costs (DIN 276 CG 700 without interim financing) accounted for 21% of cost groups 200-600.

(37) The above result includes the costs of interim financing (DIN 276 CG 760). Conventional and PPP construction costs including and excluding interim financing costs are the same as a result. This is because of two opposing effects: Although the PPP interim financing costs are generally lower due to the lower construction costs and shorter construction period, this advantage is offset by a higher interest rate for financing.

2.1.2 Qualitative aspects

(38) In terms of the construction standard, the average BKI value for new-builds was applied to ten projects, while the BKI key figures for passive house projects were used for CG 300 (building construction) and CG 400 (technical installations) in one passive house project. The construction standard of five projects is above average. This was taken into account by applying values between the BKI average and maximum values.

PPP construction standard
10 x average
1 x passive house
5 x above average

(39) The comparable BKI data enables the projects to be categorised qualitatively by comparing the main features of the construction and development, the building structure, the technical installations, the outdoor facilities and the fixtures and fittings. In the following table, five of the seven BKI comparison schools demonstrate an average building standard, while the PPP vocational school is more comparable to the two BKI vocational schools categorised as above average[10].

Example of construction standard differentiation

BKI Object data / DIN 276	CG 200: Preparation and development	CG 300: Building construction	CG 400: Technical building facilities	CG 500: Outside facilities	CG 600: Furnishings and artworks	STANDARD
No 4200-0008 **Vocational school**	Demolition of buildings Demolition of foundations	Reinforced concrete walls/ceilings/floor slab Box walls Underground car park (147 spaces)	Wood chip system Gas boiler, Heat pump with heat recovery	Exposed concrete prefab slab Traffic island	Fully furnished including technical equipment Overhead monorail	ABOVE AVERAGE
No 4200-0017 **Vocational secondary school**	Demolition of concrete paving Demolition of asphalt surfaces	Reinforced concrete basement and foundation Solid timber interior/exterior walls Cross-laminated timber ceiling Fitted cabinets and shelving	Gas boiler Underfloor heating	Asphalt concrete surface layer Bridge to existing building	Part-furnishing (tables, chairs, bins) Projectors and overhead projectors	AVERAGE
No 4200-0018 **Trade school**	Demolition of storage buildings Demolition of floor slab Felling deciduous trees	Reinforced concrete floor slab Reinforced concrete walls/ceiling Box walls	Gas boiler Heat pumps	Concrete paving Hedges	Special technical facilities Full furnishing including technical equipment	ABOVE AVERAGE
No 4200-0021 **Competence centre**		Reinforced concrete floor slab/ceiling Reinforced concrete prefab walls Partial extensive landscaping	Photovoltaic panels integrated into facade Decentralised water supply District heating Clean room	43 parking spaces Paving and asphalt surfacing Planting trees/hedges		AVERAGE
No 4200-0022 **Teaching and workshop building**	Demolition of wire mesh fence Demolition of silo Demolition of foundations Tree felling	Reinforced concrete floor slab Timber frame walls (exterior) Sand-lime brickwork (interior)	Gas condensing boiler Decentralised ventilation system with heat recovery	Fine levelling Natural stone gravel	Crystal mirror Toilet brush	AVERAGE
4200-0027 **Vocational school**	Demolition of old building	Reinforced concrete floor slab Reinforced concrete walls/ceilings	Air/ground heat exchanger Photovoltaic equipment No static heating			AVERAGE
No 4200-0030 **Vocational school**	Demolition of buildings Explosives disposal	Bored piles Reinforced concrete floor slab Reinforced concrete tank Composite masonry Extensive roof landscaping	District heating supply Borehole heat exchanger Underfloor heating Barrier system	Interlocking paving stones Seating areas Landscaping	Full furnishing including technical equipment Furnishing the cafeteria	AVERAGE
PPP Vocational school Duisburg	Clearing the site Demolition of small building	Reinforced concrete floor slab Bored piles Reinforced concrete prefab interior/exterior Extensive landscaping	Geothermal equipment Air-conditioned	Gravel paths Landscaping/flower planting	Full furnishing including technical equipment Furnishing the cafeteria	ABOVE AVERAGE

Table 2.2: Example of valuation of construction standards

(40)On construction cost certainty, see below paragraph 205 ff.

.

2.2   Construction times

a) RESULTS:

PPP-NEW construction times	PPP/BKI	GFA
Maximum	-41%	15.956
1st quartile	-35%	11.051
Median (MED)	-35%	15.770
3rd quartile	-25%	30.720
Minimum	0%	33.427
Average value (AV)	-30%	23.190
Average weighted value (AWV)	-27%	
Ø MED/AV/AWV	-30%	

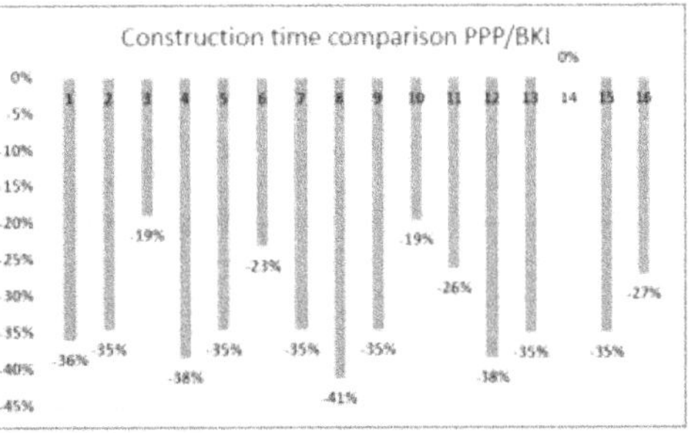

n=16

Table 2.3: PPP/BKI construction times

(41)The PPP construction time for the 16 projects is on average 30% lower than the   **PPP construction times**

BKI key figures. None of the projects analysed exceeded the comparable BKI value, ranging from 0% to -41%.

Ø 30% under BKI

(42) In the case of larger PPP projects with several locations, the efficiency of PPP construction work is not adequately reflected by a comparison of the total construction time with the conventional construction time for a building. In a PPP project with four locations, for example, the construction work was completed within the same timeframe (26 months) as a conventional project. The efficiency of the PPP service becomes obvious when considering the completed volume of construction work per month: In the case of public-private partnerships, this is €2.4 million per month, whereas for the conventional method it is €0.4 million per month. This is an advantage by a factor of 6.

High time efficiency for larger projects

b) NOTES:

(43) PPP construction time is the period from the start of construction work to the completion of the building and the subsequent start of use.

(44) For the average BKI construction time for conventional schools (25.9 months), see PPP-Schulstudie 2019, paragraph 19

(45) Efficient construction times are relevant to costs: For example, they have an impact on staff retention, the costs of construction site equipment, interim financing, interim costs or options for the previous use of old sites, etc.

(46) On PPP deadline certainty see below paragraph 210 ff.

2.3 DIN 276 CG 760 — Interim financing
2.3.1 Costs

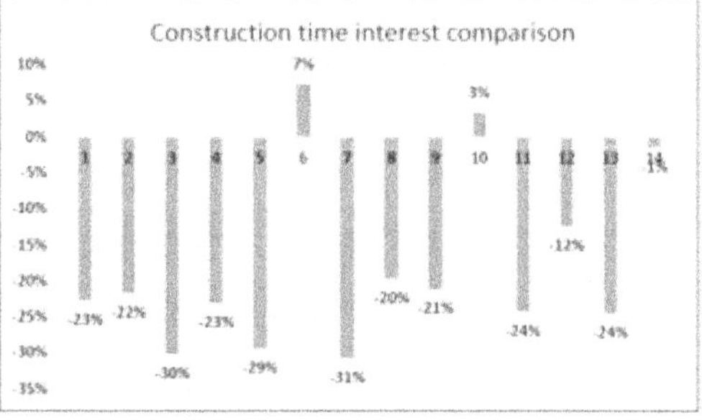

PPP NEW Interim fin. interest	PPP/CPP
Minimum	-31%
1st quartile	-24%
Median (MED)	-22%
3rd quartile	-14%
Maximum	7%
Average value (AV)	-18%
Average weighted value (AWV)	-20%
Ø MED/AV/AWV	-20%

n= 14

Table 2.4: PPP/BKI interim financing

a) RESULTS:

(47) PPP interim financing costs are on average 20% lower than the cost of conventional interim financing costs. Two PPP projects are higher than the comparable conventional value, ranging from +7% to -31%.

b) NOTES:

(48) The evaluation of the project data is based on documents provided by the PPP companies. Given the interest conditions and margin mark-ups on the reference interest rate specified in the PPP price sheets, an average margin mark-up of 63 basis points was applied to the PPP projects.

PPP additional costs for interest rate hedging

2.3.2 Qualitative aspects

(49) The reason for the higher private interest rates compared to public interest rates is, in part, the difference in credit ratings between private and public borrowers. On the other hand, the fact that the construction costs are agreed as a fixed price in public-private partnerships and therefore interest rate hedging agreements are concluded for the PPP interim financing in order to minimise the risk of interest rate increases is also relevant here. With the conventional approach, by contrast, the outflow of funds is generally financed with overdraft facilities, meaning there is generally no hedging of the interest rate risk.

Cause: PPP: Fixed construction costs price necessitates interest rate hedging

(50) The qualitative advantage of interest rate hedging is also reflected in the very

Interest rate

good PPP values for cost and deadline certainty (see below, paragraph 210).

hedging is one reason for good cost and deadline management

2.4   DIN 18960 CG 100 — Capital costs
2.4.1 Costs

a) RESULTS:

(51) PPP financing costs are on average 15% lower than conventional financing costs. The range is from 0% to -26%.

(52) On average, capital costs account for 59.2% of usage costs over the duration of the contract for public-private partnerships, 67.7% for the KGSt model and 59.0% for the BKI model (see below paragraph 155).

PPP NEW Interim fin. interest	PPP/CPP
Minimum	-31%
1st quartile	-24%
Median (MED)	-22%
3rd quartile	-14%
Maximum	7%
Average value (AV)	-18%
Average weighted value (AWV)	-20%
Ø MED/AV/AWV	-20%

n= 14

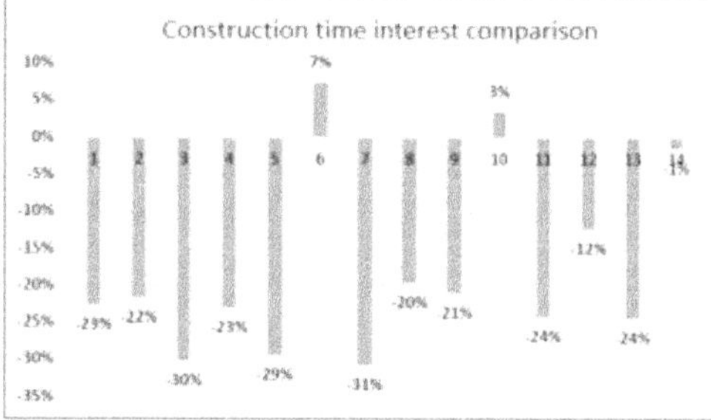

Table 2.5: Capital costs PPP/KGSt and PPP/BKI

b) NOTES:

(53) The results for capital costs are attributable to the lower construction costs, although the calculated PPP advantage is slightly reduced as a result of the higher financing interest rates.

Higher interest rates

(54) The public-private partnership and conventional procurement process capital costs are calculated based on congruent financing structures, i.e. the initial capital amount is generally repaid in full during the term of the contract (full amortisation).

Congruent financing structures

(55) Of the 16 PPP projects, the final financing for 11 projects involved forfeiting with a waiver of defence. The PPP company sold its credit claim against the local authority agreed in the PPP contract (financing of the construction costs) to a refinancing bank; the PPP local authority waived its defence to this bank under the PPP contract. As a result, the purchasing bank can negotiate loan conditions similar to those for local authorities, which the PPP company incorporates accordingly in the PPP contract. The average financing interest rates applied here were 18 basis points higher than for conventional financing.

PPP waiver of defences

(56) Five of the 16 PPP projects are classified as project financing. In this case, the refinancing bank bears the risk that the PPP local authority will impose fee reductions due to failure to adhere to the target contents of the PPP contract, thus also affecting the refinancing. Due to the associated project risks assumed, the interest rates are significantly higher in this case than for the risk-free local authority loan. Here, an average premium of 59 basis points was applied.

Project financing

(57) It should be noted that project financing incurs additional costs under other operating costs (see below under DIN 18960, CG 390).

(58) For two PPP projects, the final financing of construction costs is not part of the PPP contract. The PPP construction costs are paid by the local authority after acceptance and completion and are normally financed by borrowing.

PPP excluding final financing

2.4.2 Qualitative aspects

(59) A higher refinancing interest rate may be attributable to additional qualitative performance. This means that the PPP local authority also receives something in exchange for the project financing. Due to the assumption of project risks, the refinancing bank has a vested interest in ensuring that the PPP company

implements the project properly. Therefore, a contract review is carried out before the contract is concluded and continuous reporting obligations and control measures are put in place throughout the course of the project. For example, this is used to monitor compliance with the stipulated maintenance quality and the intended use of the maintenance reserve. In this respect, the interests of the local authority and the refinancing bank are in line, with the latter assuming an additional insurance role from the perspective of the local authority. In the conventional process, there is no corresponding quantitative approach for this.

2.5 DIN 18960 CG 200 — Property management
2.5.1 Costs

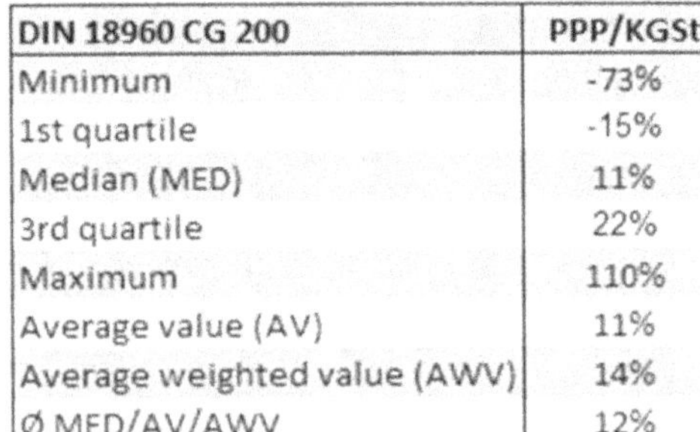

DIN 18960 CG 200	PPP/KGSt
Minimum	-73%
1st quartile	-15%
Median (MED)	11%
3rd quartile	22%
Maximum	110%
Average value (AV)	11%
Average weighted value (AWV)	14%
Ø MED/AV/AWV	12%

n=16

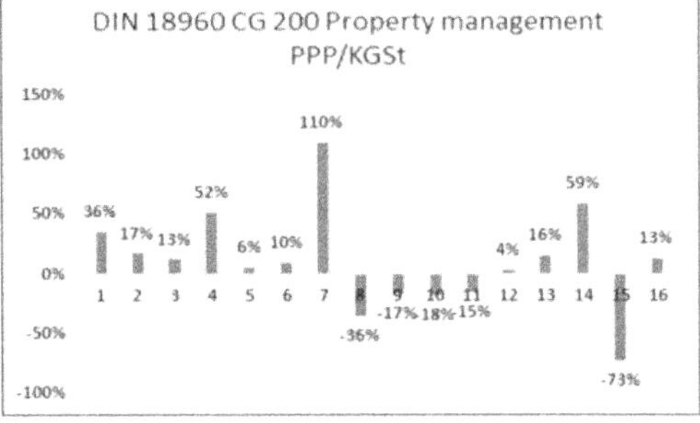

DIN 18960 CG 200	PPP/BKI
Minimum	-62%
1st quartile	-3%
Median (MED)	18%
3rd quartile	34%
Maximum	145%
Average value (AV)	23%
Average weighted value (AWV)	27%
Ø MED/AV/AWV	23%

n=16

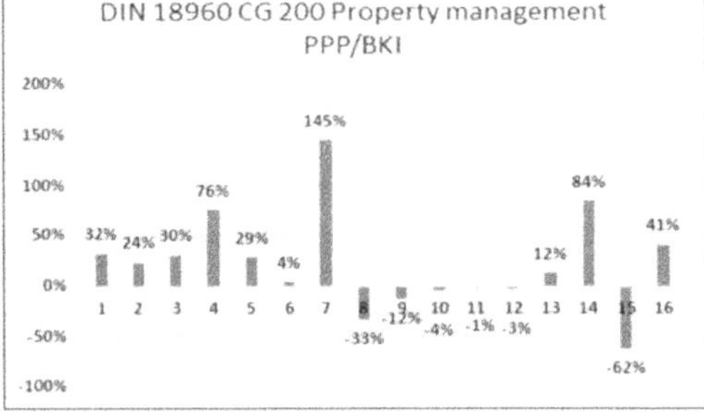

Table 2.6: Property management PPP/KGSt and PPP/BKI

a) RESULTS:

(60) Compared to the KGSt model, PPP property management costs are on average 12% higher, and 23% higher than in the BKI model. Rising project volumes result in higher values (PPP/KGSt: +14%; PPP/BKI: +27%).

PPP: Property management costs Ø +12% > KGSt Ø +23% > BKI

(61) Property management costs account for an average of 6.7% of the usage costs over the term of the contract for public-private partnerships, 6.0% for KGSt and 4.6% for BKI (see paragraph 155, for further property management costs see paragraph 122f).

+ pro rata costs from cost group 390

b) NOTES:

(62) A comparison of the 16 individual projects does not paint a cohesive picture at first glance: When compared with the KGSt model, property management costs are lower than conventional costs for five projects, and this is true for six projects when compared with the BKI model. However, it should be noted that the considerable cost differences in the cost group DIN 18960 CG 390 (other operating costs) are likely to be attributable to the fact that it includes cost items that are systematically allocated to property management costs (see below under DIN 18960 CG 390). Conversely, in projects seven and 14, elements of other cost groups could be included in the PPP property management costs.

No standardised cost allocation for PPP projects

(63) For PPP property management costs, it should be noted that PPP personnel costs are subject to VAT, while the cost of employing public employees is not. PPP projects therefore generate additional VAT income for federal, state and local authorities.

VAT on PPP personnel costs

2.5.2 Qualitative aspects

(64) Higher PPP personnel costs can be explained by the incentive and liability structures intrinsic to public-private partnerships: PPP companies have a long-term opportunity to make a profit, but in return they have a long-term contractual obligation, and must bear the cost risk and operator liability. Any deviations from the target service levels may result in fee reductions and claims for compensation. In contrast, efficient energy management allows the company to participate in cost savings in media consumption; efficient management of maintenance creates the opportunity for a pro rata payment of a credit balance remaining in the reserve account at the end of the contract. It is therefore necessary and beneficial to address this with sufficient and well-qualified personnel and a well-organised company hierarchy with project managers on site and coordinating divisional managers below the management. In the PPP company's own interest, this liability and incentive structure also requires the establishment and operation of an efficient IT-based controlling and monitoring system to manage operating processes. Naturally, the public contracting party also benefits from this.

2.6 DIN 18960 CG 300 — Operating costs

DIN 18960 CG 300	PPP/KGSt
Minimum	-19%
1st quartile	13%
Median (MED)	18%
3rd quartile	31%
Maximum	72%
Average value (AV)	21%
Average weighted value (AWV)	19%
Ø MED/AV/AWV	19%

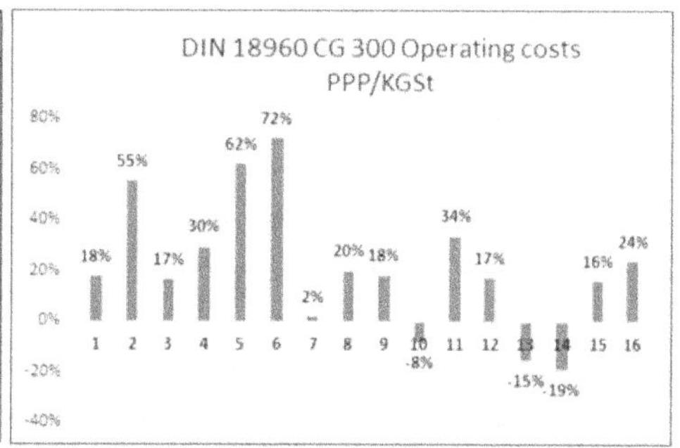

n=16

DIN 18960 CG 300	PPP/BKI
Minimum	-62%
1st quartile	-34%
Median (MED)	-29%
3rd quartile	-2%
Maximum	19%
Average value (AV)	-21%
Average weighted value (AWV)	-21%
Ø MED/AV/AWV	-24%

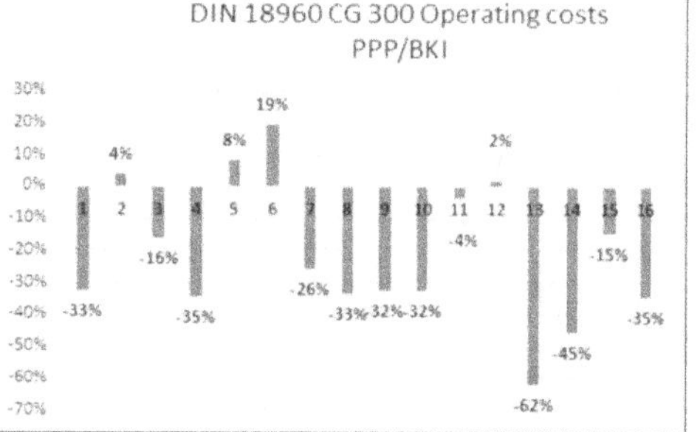

n=16

Table 2.7: Operating costs PPP/KGSt and PPP/BKI

a) RESULTS:

(65) PPP operating costs are on average 19% higher than KGSt operating costs. 13 of the 16 PPP budgets are over the KGSt key figures.

(66) Operating costs in PPP projects are on average 24% below the BKI key figures. Four of the 16 PPP projects undershot the BKI values.

(67) On average, operating costs account for 21.4% of usage costs over the duration of the contract for public-private partnerships, 17.5% for KGSt and 23.2% for BKI (see below paragraph 155).

b) NOTES:

(68) The PPP/KGSt cost difference is mainly due to higher costs for maintenance and inspection (DIN 18960 CG 350), property management (DIN 18960 CG 200) and other operating costs (DIN 18960 CG 390).

(69) PPP operating costs are lower than BKI operating costs, particularly in the cost groups utility costs (DIN 18960 CG 310) and maintenance and inspection (DIN 18960 CG 350).

PPP/BKI:
Differences in supply and maintenance

2.7 DIN 18960 CG 310 — Water, heating and electricity supply
2.7.1   Costs
2.7.1.1 Total costs for water, heating and electricity

DIN 18960 CG 310	PPP/KGSt
Minimum	-61%
1st quartile	-38%
Median (MED)	-22%
3rd quartile	-9%
Maximum	91%
Average value (AV)	-18%
Average weighted value (AWV)	-9%
Ø MED/AV/AWV	-16%

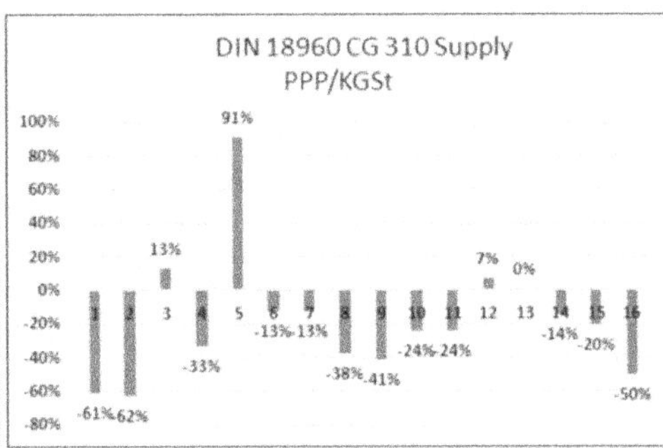

n=16

DIN 18960 CG 310	PPP/BKI
Minimum	-68%
1st quartile	-47%
Median (MED)	-39%
3rd quartile	-16%
Maximum	59%
Average value (AV)	-30%
Average weighted value (AWV)	-27%
Ø MED/AV/AWV	-32%

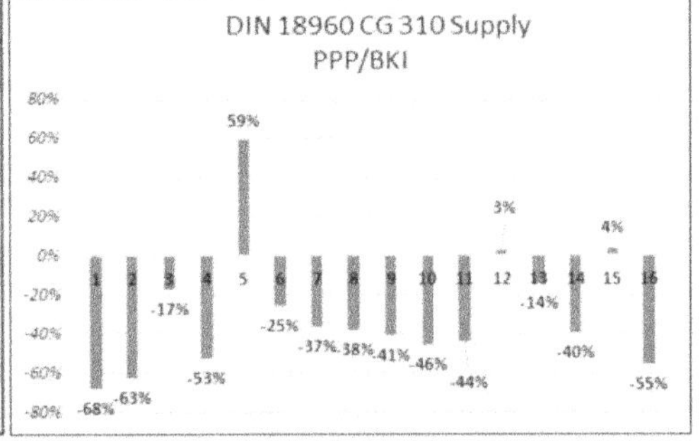

n=16

Table 2.8: Water, heating and electricity supply PPP/KGSt and PPP/BKI

a) RESULTS:

(70) PPP utilities costs are on average 16% lower than KGSt utilities costs. 12 of the 16 PPP budgets remain under the KGSt key value and four are over.

Costs for electricity, heating, water:
PPP/KGSt: Ø -16%

(71) The BKI key figures were undershot by an average of 32%. 14 of the 16 PPP budgets are under the BKI key value and two are over.

(72) Utilities costs account for an average of 5.2% of usage costs over the duration of the contract for public-private partnerships, 6.4% for KGSt and 6.5% for BKI (see below paragraph 155).

PPP/BKI: Ø -32%

b) NOTES:

(73) Of the 16 PPP projects, ten have the typical PPP contract structure, i.e. the PPP company guarantees maximum consumption volumes for water, heating and electricity[11]. If the actual consumption volumes exceed the maximum values, the PPP company bears the cost risk. If the maximum values are not reached, the two contracting parties share the savings equally; in one case, the contractor receives 65% of the savings. In one project, the PPP company assumes 100% of the surplus volume risk and the local authority is entitled to 100% of the savings. In three projects, the PPP company assumes 100% of the surplus volume risk and can also retain 100% of the savings. For two projects, the local authority bears 100% of the potential risks and rewards of surplus or insufficient amounts.

Typical PPP incentive structure:

Guaranteed max. consumption; contractor will bear the risk of excess amounts
Contractor/client share savings

(74) For every project, the same labour prices were applied for public-private partnerships and KGSt/BKI. As there is only a total cost figure for cost group 310 for BKI, the proportions of cost groups 311-313 were determined according to the KGSt distribution.

(75) In ten cases, information on the actual consumption volumes was available. Here, supply costs were initially calculated based on the actual volumes to date

and averaged over the remaining duration of the contract. In the other six cases, the costs are calculated based on the maximum consumption quantities.

DIN 18960 CG 310	PPP actual/max
Minimum	-38%
1st quartile	-17%
Median (MED)	-13%
3rd quartile	-4%
Maximum	3%
Average value (AV)	-13%
Average weighted value (AWV)	-7%
Ø MED/AV/AWV	-11%

n=10

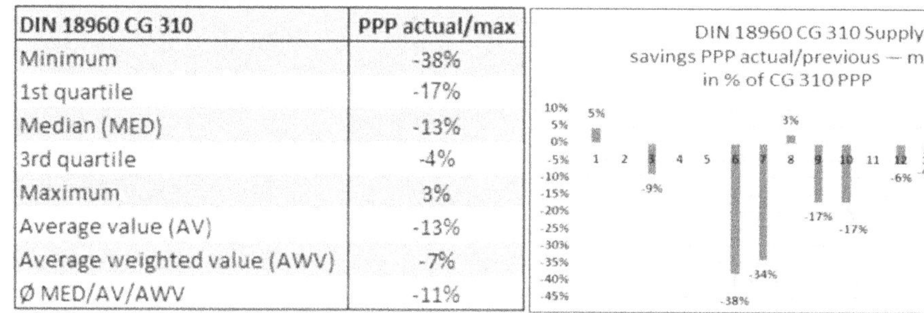

Table 2.9: Anticipated PPP savings on utilities costs (DIN 18960 CG 310) due to demand below the maximum guaranteed quantities

(76) In nine of the ten projects with data on the actual consumption volumes, these volumes fall below the maximum consumption volumes by an average of 11% relative to the supply costs (DIN 18960 CG 310) and by 0.6% relative to the usage costs (DIN 18960 CG 100-400). The actual consumption values are slightly higher than the maximum guaranteed consumption volumes in only one project. The ten projects are expected to save a total of €7.9 million in utilities costs.

Incentive system bears fruit: Savings Ø -11% (estimated €7.9 million)

2.7.1.2 Water

DIN 18960 CG 311	PPP/KGSt
Minimum	-82%
1st quartile	-56%
Median (MED)	-20%
3rd quartile	14%
Maximum	502%
Average value (AV)	12%
Average weighted value (AWV)	33%
Ø MED/AV/AWV	9%

n=16

DIN 18960 CG 312	PPP/BKI
Minimum	-82%
1st quartile	-70%
Median (MED)	-43%
3rd quartile	-27%
Maximum	378%
Average value (AV)	-10%
Average weighted value (AWV)	1%
Ø MED/AV/AWV	-17%

n=16

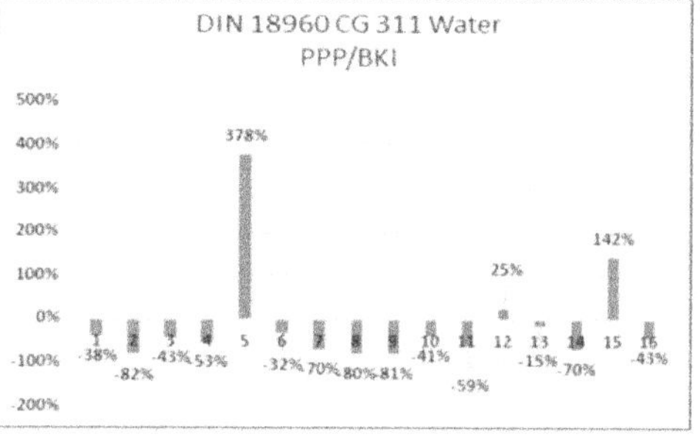

Table 2.9.1: Water supply PPP/KGSt and PPP/BKI

a) RESULTS:

(77) The PPP water costs are on average 9% higher than the KGSt model; due to an outlier project, the median value is significantly below the KGSt values (-20%),

Water supply costs PPP/KGSt: +9%

(78) On average, the BKI costs are undershot by 17%; due to the one outlier project, the median PPP costs are 43% below BKI.

PPP/BKI: -17%

b) NOTES:

(79) Ten of the 16 projects have lower costs compared to the KGSt model and 13 have lower costs than the BKI model.

Actual and max. PPP water consumption rate

(80) Data on water consumption to date was available for eight of the 16 projects, which made it possible to compare this with the guaranteed maximum

PPP actual/ PPP max: -15%

consumption quantities. To date, consumption has been on average 15% below the guaranteed maximum consumption volumes; extrapolated over the respective term of the contract, this corresponds to a reduction in consumption of 12 567 cubic metres.

DIN 18960 CG 311 Water (cbm)	PPP actual/max
Minimum	-57.472
1st quartile	-22.632
Median (MED)	-9.509
3rd quartile	-6.793
Maximum	36.172
Average value (AV)	-13.483
Average weighted value (AWV)	-14.709
Ø MED/AV/AWV	-12.567

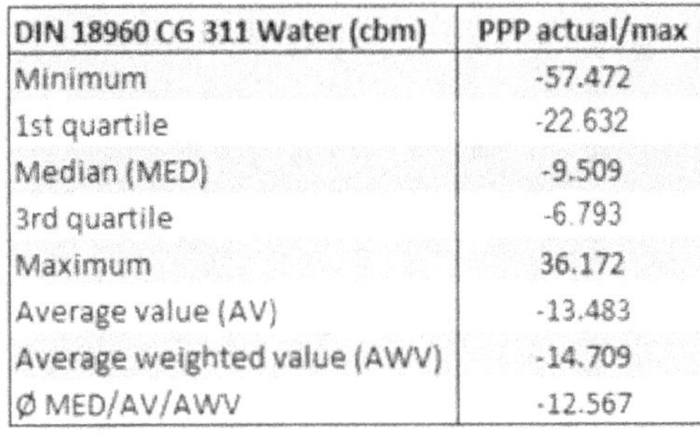

n=8

DIN 18960 CG 311 Water (cbm)	PPP actual/max
Minimum	-70%
1st quartile	-24%
Median (MED)	-18%
3rd quartile	-11%
Maximum	43%
Average value (AV)	-16%
Average weighted value (AWV)	-10%
Ø MED/AV/AWV	-15%

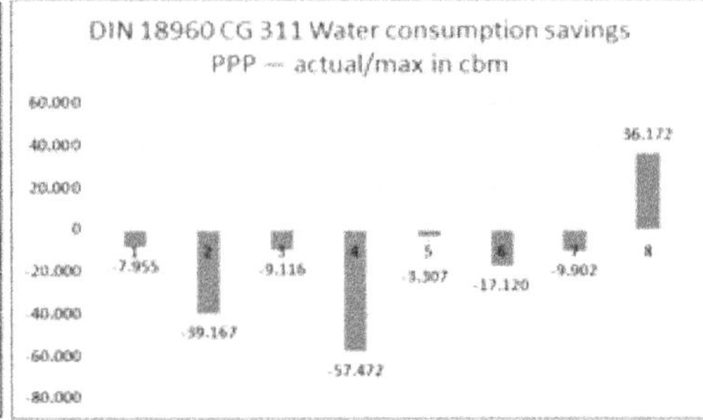

n=8

Table 2.9.2: PPP: Previous/projected actual and guaranteed max. water consumption

2.7.1.3 Heating

DIN 18960 CG 312	PPP/KGST
Minimum	-83%
1st quartile	-50%
Median (MED)	-34%
3rd quartile	-17%
Maximum	4%
Average value (AV)	-33%
Average weighted value (AWV)	-27%
Ø MED/AV/AWV	-31%

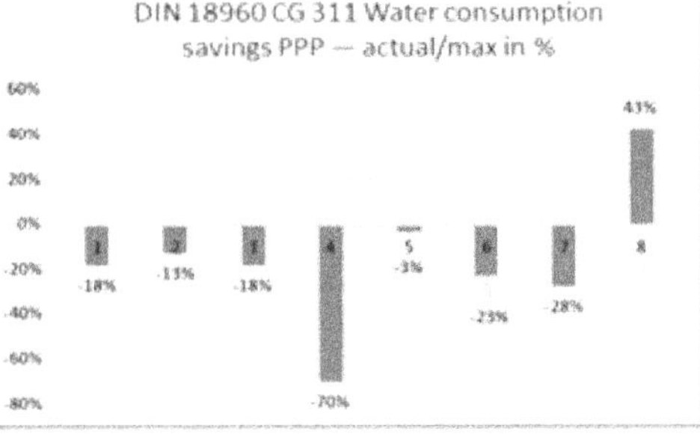

n=16

DIN 18960 CG 312	PPP/BKI
Minimum	-83%
1st quartile	-56%
Median (MED)	-48%
3rd quartile	-32%
Maximum	-7%
Average value (AV)	-42%
Average weighted value (AWV)	-38%
Ø MED/AV/AWV	-43%

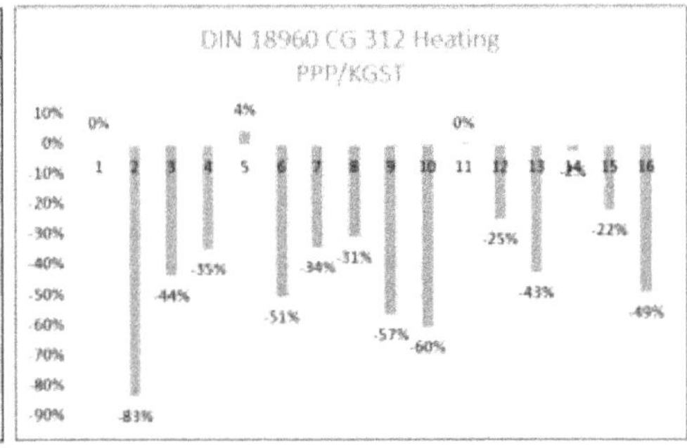
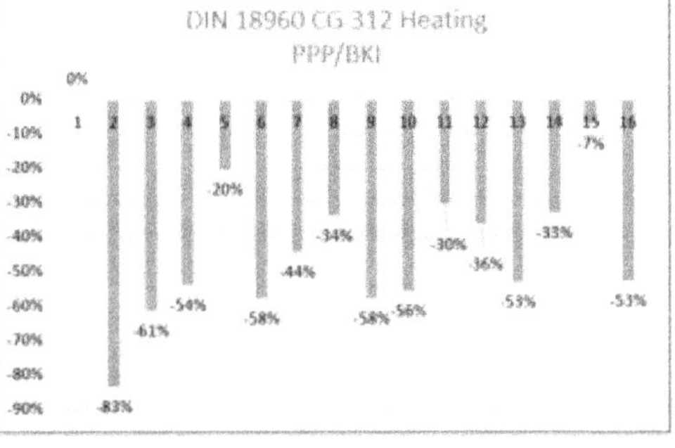

n=16

Table 2.9.3: Heating supply PPP/KGSt and PPP/BKI

a) RESULTS:

(81) The PPP heating costs are on average 31% lower than the KGSt model; the BKI costs are undershot by 43% on average.

Heating supply costs
PPP/KGSt: -31%
PPP/BKI: -43%

b) NOTES:

(82) 14 of the 16 PPP projects have lower costs compared to the KGSt model and 15

have lower costs than the BKI model.

(83) Data on heat consumption to date was available for nine of the 16 projects, which made it possible to compare this with the guaranteed maximum consumption quantities. To date, consumption has been on average 32% below the guaranteed maximum consumption volumes — only in one project are the actual values slightly above the agreed maximum value. Extrapolated over the respective duration of the contract, this results in a total reduction in consumption of 11 119 317 kWh. With average CO_2 emissions of 201 g per kWh of natural gas[12], this would mean CO_2 savings of 20 100 tonnes over 25 years.

Heat consumption

Actual and maximum PPP heating consumption rate PPP actual/ PPP max: Ø -15%

CO_2 savings: 20 100 tonnes

DIN 18960 CG 312 Heating (kWh)	PPP actual/max
Minimum	-39.867.392
1st quartile	-12.931.893
Median (MED)	-7.774.015
3rd quartile	-1.401.740
Maximum	684.632
Average value (AV)	-10.358.763
Average weighted value (AWV)	-15.225.174
Ø MED/AV/AWV	-11.119.317

n=9

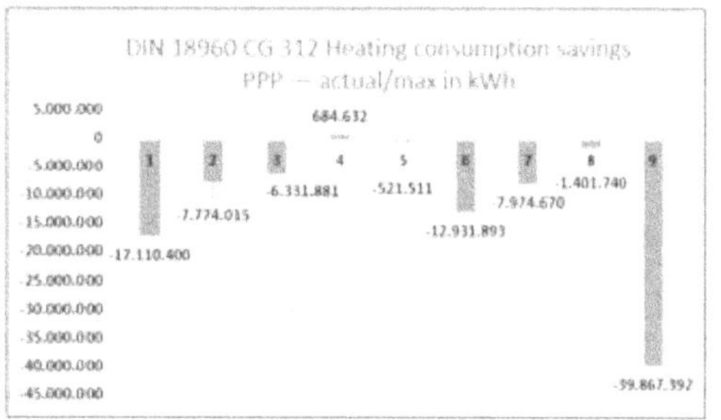

DIN 18960 CG 312 Heating (kWh)	PPP actual/max
Minimum	-61%
1st quartile	-42%
Median (MED)	-35%
3rd quartile	-22%
Maximum	5%
Average value (AV)	-32%
Average weighted value (AWV)	-29%
Ø MED/AV/AWV	-32%

n=9

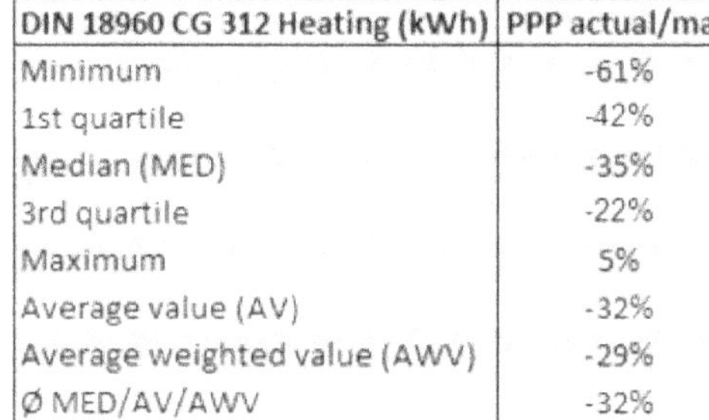
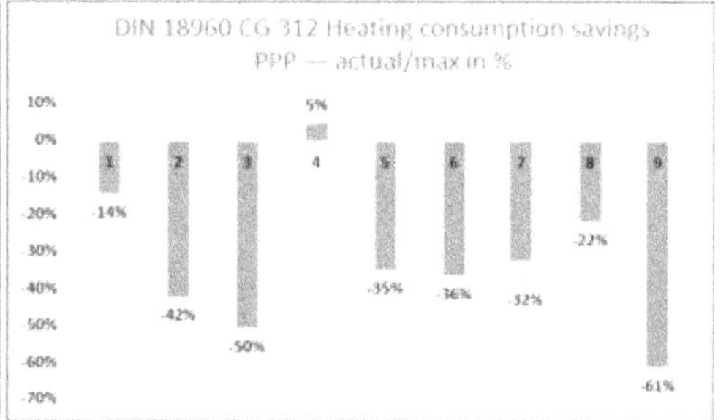

Table 2.9.4: PPP: Previous/projected actual and guaranteed max. heating consumption

2.7.1.4 Electricity

a) RESULTS:

(84) The PPP electricity costs are on average 26% higher than the KGSt model; due to two outlier projects, the PPP median value is only 17% higher than for KGSt.

(85) The BKI costs are exceeded by 4% on average; the median PPP electricity costs are 11% below BKI.

Electricity supply costs

PPP/KGSt: +26%

PPP/BKI: +4%

DIN 18960 CG 313	PPP/KGSt
Minimum	-59%
1st quartile	-32%
Median (MED)	17%
3rd quartile	24%
Maximum	209%
Average value (AV)	14%
Average weighted value (AWV)	46%
Ø MED/AV/AWV	26%

n=16

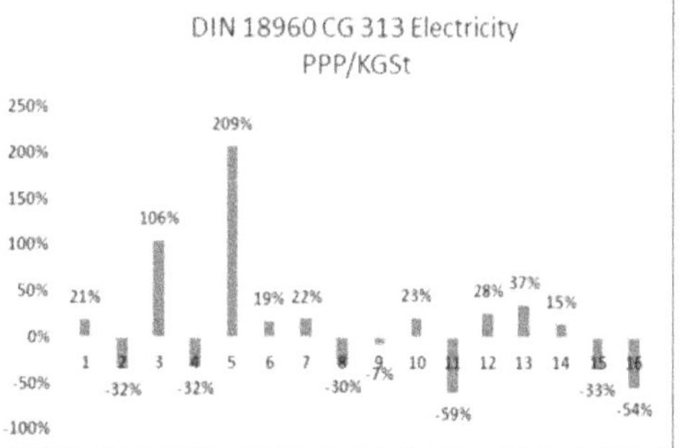

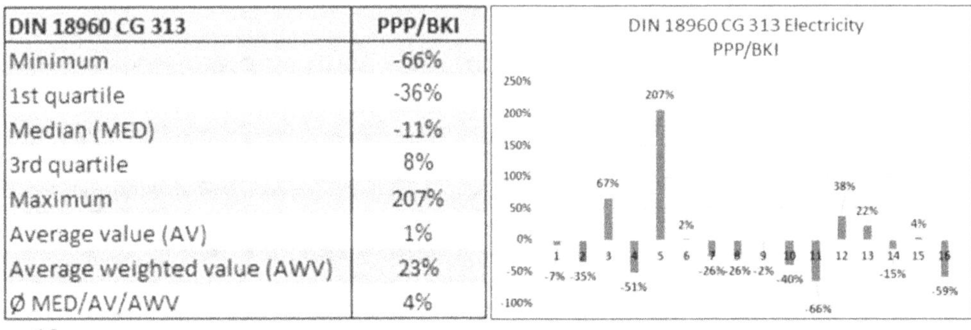

DIN 18960 CG 313	PPP/BKI
Minimum	-66%
1st quartile	-36%
Median (MED)	-11%
3rd quartile	8%
Maximum	207%
Average value (AV)	1%
Average weighted value (AWV)	23%
Ø MED/AV/AWV	4%

n=16

Table 2.9.5: Electricity supply PPP/KGSt and PPP/BKI

b) NOTES:

(86) Compared to the KGSt model, nine of the 16 PPP projects have higher costs. The figure for the BKI model is six. Two of the projects are outliers.

(87) Data on electricity consumption to date was available for ten of the 16 projects, which made it possible to compare this with the guaranteed maximum consumption quantities. To date, consumption has been on average 3% below the guaranteed maximum consumption volumes; extrapolated over the respective term of the contract, this corresponds to a reduction in consumption of 1 153 150 kWh. With average CO_2 emissions of 420 g per kWh of electricity[13], the ten projects will save 4 840 tonnes of CO_2 over 25 years.

Actual and maximum PPP electricity consumption rate

PPP actual/ PPP max: -3%

CO_2 savings: 4 840 tonnes

DIN 18960 CG 313 Electricity (kWh)	PPP actual/max
Minimum	-5.841.370
1st quartile	-1.615.940
Median (MED)	-707.148
3rd quartile	808.963
Maximum	2.061.023
Average value (AV)	-917.384
Average weighted value (AWV)	-1.834.919
Ø MED/AV/AWV	-1.153.150

n=10

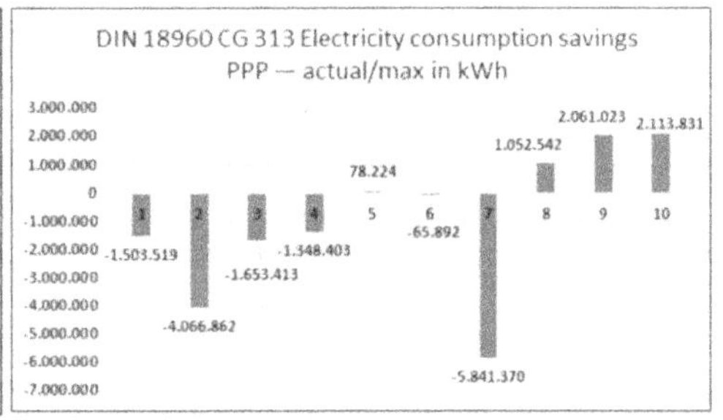

DIN 18960 CG 313 Electricity (kWh)	PPP actual/max
Minimum	-26%
1st quartile	-16%
Median (MED)	-4%
3rd quartile	6%
Maximum	46%
Average value (AV)	-1%
Average weighted value (AWV)	-4%
Ø MED/AV/AWV	-3%

n=10

Table 2.9.6: PPP: Previous/projected actual and guaranteed max. electricity consumption

2.7.2   Qualitative aspects

2.7.2.1 Heat consumption comparison

(88) Information on the guaranteed maximum consumption volumes was available for 12 of the 16 new-build projects, and for nine projects information was also available on the actual consumption volumes from the previous operating period. The guaranteed and actual consumption values can then be compared with the guide and average values as per VDI 3807 sheet 2 and KGSt consumption data.

European comparative PPP study
Heat consumption comparison

Project no.	1	2	3	4	5	6	7	8	9	10	11	12	
Building type	VOCe xSH	GEN+ SH	AB	GEN+ SH	VOCe xSH	VOCe xSH	GEN+ SH	GEN+ SH	AB	AB / VOCex SH	GEN+ SH	GENe xSH	
Dataset		2009-2019	2006-2020	2016-2020	2014-2020	2014-2020	2010-2019		2013-2019	2013-2019	2012-2019		Ø
Maximum guaranteed kWh/sqm	15	74	66	60	39	38	48	87	56	40	78	42	
Actual kWh/sqm GFA		63	38	30	39	25	30		38	31	30		
Difference actual/max		-14%	-42%	-50%	2%	-35%	-36%		-32%	-22%	-61%		-32%
VDI 3807 sheet 2 reference value	52	68	55	68	52	52	68	68	55	54,0	68	65	
Difference to max	-71%	9%	20%	-12%	-26%	-27%	-30%	0,29	3%	-26%	15%	-35%	-8%
Difference from actual		-7%	-30%	-56%	-25%	-53%	-55%		-31%	-42%	-55%		-38%
VDI 3807 sheet 2 average	82	99	80	99	82	82	99	99	80	81	99	95	
Difference to max	-81%	-25%	-17%	-39%	-56%	-54%	-52%	-12%	-29%	-50%	-21%	-56%	-38%
Difference from actual		-36%	-52%	-70%	-52%	-70%	-69%		-52%	-61%	-69%		-58%
KGSt — Actual	80	99	70	99	80	80	99	99	70	73	99	89	
Difference to max	-81%	-25%	-6%	-39%	-59%	-53%	-52%	-12%	-20%	-45%	-21%	-53%	-36%
Difference from actual		-36%	-45%	-70%	-51%	-69%	-69%		-46%	-57%	-69%		-56%

Key: VOCexSH = vocational school without sports hall; GEN+/exSH = general education school with/without sports hall; AB = administrative building

Comparison actual and maximum PPP heating consumption rate

with VDI guideline value:
Ø -8% (max)/
Ø -38% (actual)

with VDI average:
Ø -38% (max)/
Ø -58% (actual)

with KGSt:
Ø -36% (max)/
Ø -56% (actual)

Table 2.10: Heat consumption comparison

(89) With regard to heat consumption, the actual values fall below the maximum consumption values guaranteed when the contract was concluded by an average of 32%. The actual values are only slightly above the agreed maximum value in one project.

(90) The agreed maximum values are on average 8% lower than the VDI 3807 guideline value; for five projects they are higher. The actual values are all below the VDI guideline value, on average 38%.

(91) All maximum and actual values fall below the VDI 3807 mean value, with the maximum values averaging 38% and the actual values 58% below the VDI mean value.

(92) The comparable values for the KGSt properties are 36% lower than the maximum values and 56% lower than the actual values.

2.7.2.2 Reduction of transmission heat loss through contractual incentives

(93) As part of a master's thesis supervised by the Technical University of Karlsruhe/Karlsruhe Institute of Technology (Prof. Dr Lennerts) and the Mainz University of Applied Sciences (Prof. Dr Bogenstätter), the relevant energy data from the tender, planning and contractual documents, including the EnEV energy performance certificates of ten PPP projects, were analysed in depth.

Master's thesis PPP und Energieeffizienz, Karlsruhe/Mainz 2022

(94) A link was established between the reduction in transmission heat losses and the contractual arrangement for the PPP company to participate in the heat consumption savings.

EnEV requirements for the reduction of transmission heat loss

(95) In one of the ten projects, the PPP company only bears the risk of the actual heating consumption volumes exceeding the contractually guaranteed quantities; 100% of the savings benefit the local authority party to the contract. As a result of the PPP planning, the transmission heat loss was 36% lower than the EnEV requirement.

(96) In the second case group, the PPP company bears the surplus volume risk but receives a 50% share of the savings. In this case, the reduction in transmission heat loss is 45% on average compared to the EnEV requirement.

(97) In the third case group, the PPP company bears 100% of the surplus volume risk as well as the opportunity for savings. In both projects, transmission heat loss

was reduced by 69% in accordance with EnEV requirements. Evidently, planning to optimise transmission heat loss is even more strongly incentivised when the PPP company will share in subsequent savings.

European comparative PPP study here: Relationship between PPP planning for reducing transmission heat losses and contractual incentives								
No.	Construction project	EnEV status	**Contractual provision**		**Transmission heat loss**			
			Contractor risk actual > max	Contractor prospects actual < max**	Planned	EnEV specification [W/m²K]		
1	New construction	2004	100%	0%	0,56	0,87	-36%	-36%
2	New construction	2007	100%	50%	0,25	0,7	-64%	
3	New construction	2004	100%	50%	0,4	0,76	-47%	
4	Renovation	2004	100%	50%	0,43	0,91	-53%	
5	New construction	2004	100%	50%	0,55	0,75	-27%	-45%
6	Renovation	2004	100%	50%	0,46	0,68	-32%	
7	New construction	2004	100%	50%	0,47	0,84	-44%	
8	New construction	2014	100%	50%	0.17 (wall) 1.0 (window)	0.35 (walls) 1.9 (windows)	-49%	
9	New construction/renovation	2007	100%	100%	0,42	1,34	-69%	-69%
10	New construction	2007	100%	100%	0,39	1,29	-70%	

* Contractor is liable for the cost risk if the guaranteed usage quantities are exceeded

** Contractor will receive a portion of the savings if consumption is below the guaranteed quantities

Source: Vöst, Sebastian (2022). Energieeffizienz bei PPP-Projekten, Masterarbeit TU KIT Karlsruhe/HS Mainz

Table 2.11: Reduction of transmission heat losses and contractual PPP incentive system

Correlation to contractor savings participation:

PPP -36%< EnEV with 0% contractor participation

PPP -45%< EnEV with 50% contractor participation

PPP -69%< EnEV with 100% contractor participation

2.8   DIN 18960 CG 330/40 — Cleaning
2.8.1 Costs

DIN 18960 CG 330	PPP/KGST
Minimum	-34%
1st quartile	-11%
Median (MED)	6%
3rd quartile	22%
Maximum	58%
Average value (AV)	7%
Average weighted value (AWV)	2%
Ø MED/AV/AWV	5%

n=16

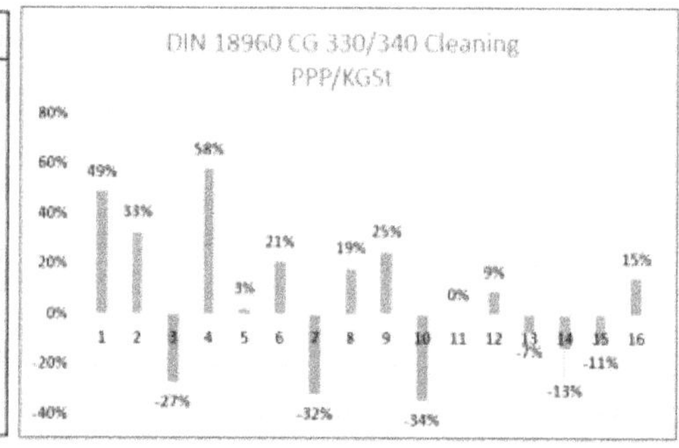

DIN 18960 CG 330	PPP/BKI
Minimum	-27%
1st quartile	-13%
Median (MED)	12%
3rd quartile	26%
Maximum	76%
Average value (AV)	13%
Average weighted value (AWV)	2%
Ø MED/AV/AWV	9%

n=16

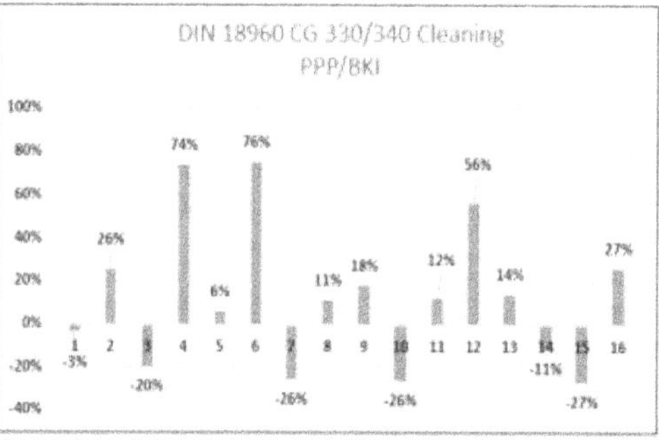

Table 2.12: Cleaning costs PPP/KGSt and PPP/BKI

a) RESULTS:

(98) PPP cleaning costs are on average 5% higher than KGSt cleaning costs. Ten of the 16 PPP budgets are over the KGSt value.

Cleaning costs: PPP/KGSt: Ø +5%

(99) The BKI key figures were exceeded by an average of 9%. Ten of the 16 PPP

PPP/BKI: Ø +9%

budgets are over the BKI value.

(100) Cleaning costs account for an average of 8.5% of the usage costs over the term of the contract for public-private partnerships, 8.2% for KGSt and 6.8% for BKI (see below paragraph 155).

b) NOTES:

(101) Feedback received to date on the deadline and cost certainty questionnaire indicates that fees have only been reduced due to complaints about cleaning services in rare cases and only to a minor extent.

To date no or very few complaints

2.8.2 Qualitative aspects

(102) A detailed evaluation of the cleaning performance was carried out for one of the 16 projects. Here, 38 different cleaning services were specified in the PPP contract for the 624 rooms divided into 17 room groups with a cleaning area of 25 225 m² and a cleaning interval is specified for each cleaning service.

Example for evaluation of cleaning

New vocational college central Duisburg Cleaning and annual cleaning area in m²		Working days	1 x week	1-2 x week	1x month	Total
A1	Classrooms	2.320				
A2	Specialist rooms	4.755				
C	Office rooms	1.930				
D	Teaching resource rooms		1.325			
E	Copier rooms		240			
F1	Circulation areas	6.951				
H1	School sanitary facilities	572				
H2	Sport sanitary facilities	485				
I	Assembly hall			350		
J	Social rooms	340				
L	Storage, archives				771	
M1	Sports areas	1 864				
N	Platform			630		
O	Building services				1 192	
P5	Outdoor area			1 500		
	Cleaning area	19.217	1 565	2 480	1 963	25 225
	Cleaning days/year	200	40	60	12	162
	Annual cleaning area	3 843 400	62 600	148 800	23 556	4 078 356

Table 2.13: Example PPP cleaning area

(103) When compared with DIN 77400, it is apparent that the PPP cleaning intervals correspond to DIN 77400 for 58% of the cleaning services, while 30% include additional or differently categorised cleaning services that do not exist in DIN 77400. The cleaning intervals are 6% more or less intensive respectively than in DIN 77400.

Cleaning costs	Max	PPP	KGSt	Difference PPP-KGSt	
Cleaning area in m²	25.225	25.225	18.588	6.637	36%
Annual cleaning area (AC	5.045.000	4.078.356	2.269.845	1.808.511	80%
Cleaning frequency (days	200	162	122	40	32%
Cleaning costs	472.533 €	472.533 €	299.800 €	172.733 €	58%
Costs/m² ACA	0.09 €	0.116 €	0.132 €	-0.016 €	-12%

If the cleaning intensity is factored in, the financial disadvantage becomes a PPP advantage

Table 2.14: Comparison of (annual) cleaning area PPP/KGSt

(104) Overall, the average cleaning interval for a PPP project is 162 days per year. The costs per m2 of annual cleaning area are €0.116/m2, which are therefore 6% lower than the comparable KGSt data (€0.124/m2 of annual cleaning area). When the cleaning intensity is factored in, the financial disadvantage becomes an advantage.

(105) Whether and to what extent such results can be seen in the other projects could not be verified in the work carried out to date.

Cleaning services comparison — Room groups

		Room type	Classrooms			Specialist rooms			Office rooms		
	PPP project/DIN 77400 (Extract)	Number/m²	31	2320		64	4755		77	1930	
	Description of cleaning service		A1	DIN	Cf.	A2	DIN	Cf.	C	DIN	Cf.
	Floors										
1	Skirting boards, damp wipe		1	6xY	⊕	1	6xY	⊕	1	6xY	⊕
2	Beverage area, wet cleaning (scrubbing)		2	-	+	2	-	+	1	-	+
3	Footprints, heel marks (scrubbing)		2	1xW	⊕	2	1xW	⊕	1	1xW	∅
4	Removal of lint/fluff		WD	1xW	⊕	WD	-	+	WD	1xW	⊕
5	Stains on textile flooring (2m²), stain remover		WD	A/R	⊕	WD	A/R	⊕	WD	A/R	⊕
6	Professional and intensive cleaning and maintenance of floors (as required: sweeping, damp/wet mopping, vacuuming, brush-vacuuming and brushing, cleaner, polishing as required, circulation areas at least 1x/week).		2	2xW	∅	2	2xW	∅	2	2xW	∅
7	Professionally eliminate major soiling		WD	1xW	⊕	WD	1xW	⊕	WD	1xW	⊕
	Fixtures and fittings										
8	Damp wipe fingerprints from doors/partitions/windows		WD	1xW	⊕	WD	1xW	⊕	WD	1xW	⊕
9	Full wet clean of doors, door frames, exterior doors		M1	-	+	M1	-	+	M1	-	+
10	Other glass surfaces, remove fingerprints on both sides		A/R	-	+	A/R	-	+	A/R	-	+
11	Sweep and vacuum doormats, brush mats (outside)		-	-	∅	-	-	∅	-	-	∅
12	Sweep and vacuum doormats, brush mats (inside)		-	-	∅	-	-	∅	-	-	∅
13	Damp clean fingerprints from light switches and sockets		1	1xW	∅	1	-	∅	1	-	+
14	Remove cobwebs		A/R	6xY	∅	A/R	A/R	+	A/R	A/R	∅
15	Damp wipe light fittings < 1.8 m		Q1	-	+	Q1	-	+	Q1	-	+
16	Interior doors and partitions (full-surface)		Q1	-	+	Q1	-	+	Q1	-	+
17	Fingerprints on furnishings (tables, display cabinets, cupboards) <1.8 m		WD	2xM	⊕	WD	-	+	WD	-	+
18	Damp wipe desks, school desks, etc. (if clear)		1	2xW	⊖	1	2xW	⊖	1	1xW	∅
19	Telephones, table lamps, office equipment		M1	-	+	M1	-	+	M1	1xW	∅
20	Damp clean waste bins		M1	A/R	∅	M1	-	+	M1	-	+
21	Waste bins/ashtrays/recycling bins/ensure waste is sorted		WD	1xW	⊕	WD	2xW	⊕	WD	1xW	⊕
22	Damp clean waste bin/hygiene bin/shredder at least 1x/month		M1	-	+	M1	1xM	∅	M1	1xM	∅
23	Damp wipe ashtray		WD	-	+	WD	-	+	WD	-	+
24	Damp wipe display cabinets, coat racks, umbrella stands, fire extinguishers, etc.		1	1xW	∅	1	1xW	∅	1	1xW	∅
25	Damp wipe seats, chairs, etc.		M1	1xM	∅	M1	1xM	∅	M1	2xY	⊕
26	Damp wipe horizontal and vertical surfaces, cupboards, tables, shelving and racks < 2.15 m		M1	-	+	M1	1xM	∅	M1	1xM	∅
27	Damp wipe radiators		Q1	6xY	⊖	Q1	6xY	⊖	Q1	6xY	⊖
28	Damp wipe windowsills		1	2xM	⊕	1	2xM	⊕	1	2xM	⊕
29	Damp wipe wash basins, taps, counters, mirrors		WD	2xW	⊕	WD	2xW	⊕	-	1xW	⊖
30	Damp wipe wall tiles in splash zone		WD	-	+	WD	2xW	⊕	-	1xW	⊖
31	Damp wipe other tiled walls		-	-	∅	-	2xW	∅	-	1xW	⊖
32	Damp wipe hand dryers, paper holders, soap dispensers		WD	-	+	WD	-	+	-	-	∅
33	Damp wipe toilet brush handle and holder		-	-	∅	-	-	∅	-	-	∅
34	Wet clean urinals, toilet bowls (inside and out)		-	-	∅	-	-	∅	-	-	∅
35	Wet clean toilet seat including lid and rim area		-	-	∅	-	-	∅	-	-	∅
36	Refill toilet paper, soap and paper towel dispensers		WD	A/R	⊕	WD	A/R	⊕	-	A/R	∅
37	Damp wipe handrails		-	-	∅	-	-	∅	-	-	∅
38	Damp wipe stair railings		-	-	∅	-	-	∅	-	-	∅

	Evaluation	Total		Classrooms		Specialist rooms		Office rooms	
⊕	PPP cleaning cycle more intensive than DIN	44	0,06	⊕	11	⊕	10	⊕	8
+	Differing/additional PPP cleaning service	204	30%	+	11	+	12	+	9
⊖	PPP cleaning intervals shorter than DIN	40	0,06	⊖	2	⊖	2	⊖	4
∅	Same as DIN/not evaluable	396	58%	∅	14	∅	14	∅	17

Key: WD = working day / 1 = last cleaning day of the week / 2.5 = every 2nd cleaning day / 2 = 2 days per week / 15 = on the 15th of every month / 15m = on the 15th of every month and on the last cleaning day of the month / B = cleaning according to prior arrangement / A/R = cleaning as required / M1 = last cleaning day of the month / M2 = 2x per month / Q1 = last cleaning day of the quarter / H1 = last cleaning day of the half-year / Y1 = last cleaning day of the year / 1xW = 1 x per week / 2xW = 2 x per week / 5xW = 5 x per week / 2xY = 2 x per year / 6xY = 6 x per year / 1xM = 1 x per month / 2xM = 2 x per month

Table 2.15: Comparison cleaning PPP/DIN 77400

2.9   DIN 18960 CG 350 — Maintenance and inspection
2.9.1 Costs

a) RESULTS:

DIN 18960 CG 350	PPP/KGST
Minimum	386%
1st quartile	591%
Median (MED)	655%
3rd quartile	736%
Maximum	1663%
Average value (AV)	747%
Average weighted value (AWV)	653%
Ø MED/AV/AWV	685%

n=10

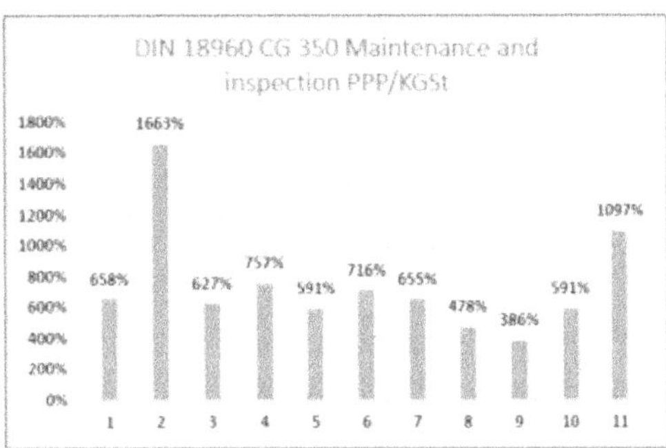

DIN 18960 CG 350	PPP/BKI
Minimum	-68%
1st quartile	-49%
Median (MED)	-27%
3rd quartile	-5%
Maximum	88%
Average value (AV)	-19%
Average weighted value (AWV)	-17%
Ø MED/AV/AWV	-21%

n=10

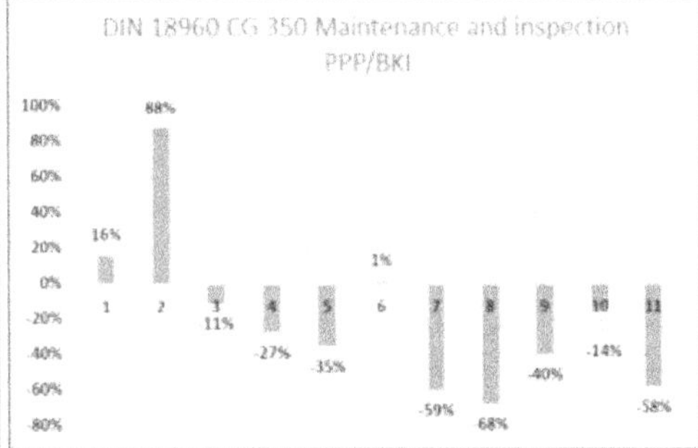

Table 2.16: Maintenance and inspection PPP/KGSt and PPP/BKI

(106) The PPP maintenance and inspection costs are on average 685% higher than the KGSt values. All PPP projects where the costs in this cost group are recorded are higher than the KGSt key figure.

> Costs of maintenance and inspection: PPP/KGSt: Ø +685%

(107) In contrast, on average the PPP costs are 21% lower than the BKI key figures. Three of the ten PPP projects with costs in this cost group are higher than the BKI key figure and seven are lower.

> PPP/BKI: Ø -21%

(108) On average, maintenance and inspection account for 6.1% of usage costs over the duration of the contract for public-private partnerships, 0.7% for KGSt and 7.4% for BKI (see below paragraph 155).

b) NOTES:

(109) The KGSt actual values are significantly lower than the PPP costs, while the PPP inspection and maintenance costs are lower than the BKI key figures.

> KGSt actual value significantly under PPP and BKI

(110) The KGSt actual values differ considerably from both the PPP and the BKI figures. When comparing usage costs, the maintenance and inspection budget is the fourth largest single item for public-private partnerships at 6.1% after capital costs (59.2%), repairs (12.6%) and cleaning (8.5%). In the BKI key figures, the cost group for maintenance and inspection is in third place at 7.4%, behind capital costs (59.0%), repair costs (13.1%) and ahead of cleaning costs (6.8%).

> Substantially different priorities for PPP and BKI

(111) The KGSt target value for the appropriate allocation of maintenance budgets does not specify a separate reference value for maintenance and inspection. However, the overall KGSt target value for maintenance was confirmed in the PPP school study by means of part-specific maintenance calculations according to the BNB guidelines (evaluation system for sustainable construction), PPP reference property and specialist literature (see PPP-Schulstudie 2019, paragraph 68 ff.; Annex A).

> KGSt target values confirmed by BNB guidelines (evaluation system for sustainable construction and specialist literature as part of the PPP school study (2019)

2.9.2 Qualitative aspects

(112) Strict adherence to all inspection and maintenance dates, including complete documentation of these, is given significantly higher priority in public-private partnerships than in conventional practice. There are significantly fewer staff and budget funds available, which means that maintenance can only ever reach the standard of a contingency plan.

KGSt: Contingency plan

(113) The reason for these significant differences may lie in the contractual incentive structures: If the standards defined in the service levels are not met, remuneration may be reduced. The PPP company must also insure itself against the risk of operator liability (see Scheidt and Konstantin (2020). Betreiberhaftung bei der konventionellen und PPP-Instandhaltung. Münster/Mainz). For this reason, PPP maintenance follows a preventive strategy focused on quality management.

PPP: Preventative strategy: Service levels, risk of fee reductions, operator liability

2.10   DIN 18960 CG 370 — Taxes and insurance

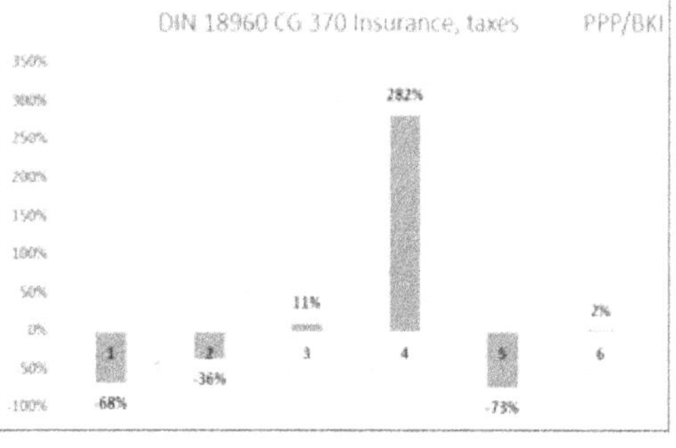

DIN 18960 CG 370	PPP/KGST
Minimum	1%
1st quartile	47%
Median (MED)	178%
3rd quartile	297%
Maximum	558%
Average value (AV)	209%
Average weighted value (AWV)	149%
Ø MED/AV/AWV	179%

n=6

DIN 18960 CG 370	PPP/BKI
Minimum	-73%
1st quartile	-60%
Median (MED)	-17%
3rd quartile	9%
Maximum	282%
Average value (AV)	20%
Average weighted value (AWV)	-24%
Ø MED/AV/AWV	-7%

n=6

Table 2.17: Taxes and insurance PPP/KGSt and PPP/BKI

a) RESULTS:

(114) The PPP costs for taxes and insurance are on average 179% higher than the KGSt values. All six PPP projects with costs in this cost group are higher than the KGSt key figure.

Costs of taxes pp.: PPP/KGSt: Ø +179%

(115) In contrast, on average the PPP costs are 7% lower than the BKI key figures. Three of the six PPP projects with costs in this cost group are higher than the BKI key figure.

PPP/BKI:Ø -7%

(116) On average, this cost group accounts for 0.9% of usage costs over the duration of the contract for public-private partnerships, 0.4% for the KGSt model and 1.1% for the BKI model (see below paragraph 155).

b) NOTES:

(117) Costs for taxes and insurance are only reported separately in six of the 16 PPP projects. For other projects, these costs may be included in the cost group DIN 18960 CG 200 (property management) or CG 390 (other operating costs).

2.11   DIN 18960 CG 390 — Other operating costs

a) RESULTS:

DIN 18960 CG 390	PPP/KGST
Minimum	-37%
1st quartile	113%
Median (MED)	317%
3rd quartile	1394%
Maximum	1641%
Average value (AV)	690%
Average weighted value (AWV)	359%
Ø MED/AV/AWV	455%

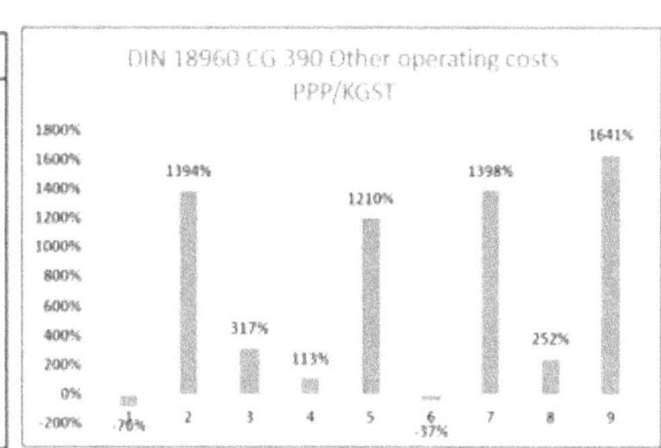

n=9

DIN 18960 CG 390	PPP/BKI
Minimum	-70%
1st quartile	37%
Median (MED)	90%
3rd quartile	5186%
Maximum	15767%
Average value (AV)	3289%
Average weighted value (AWV)	1600%
Ø MED/AV/AWV	1660%

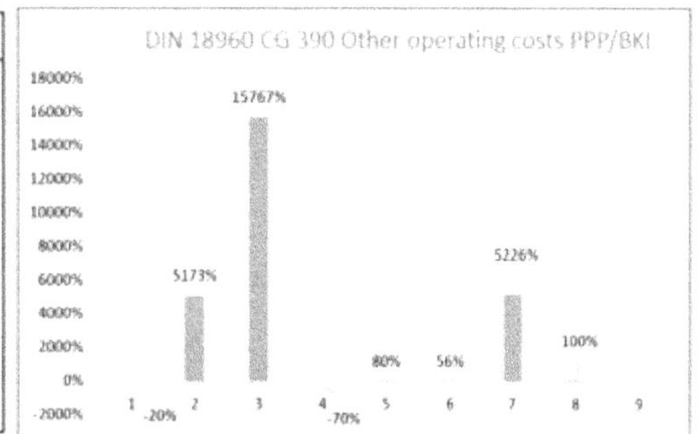

n=8

Table 2.18: Other operating costs PPP/KGSt and PPP/BKI

(118) Other PPP maintenance costs are on average 455% higher than the KGSt values. Eight of the nine PPP projects with costs in this cost group are higher than the KGSt key figure.

Other operating costs:
PPP/KGSt: Ø +455%

(119) Other PPP operating costs are on average 1660% higher than the BKI key figures. One PPP project with reported costs in this cost group came in below the BKI key figure.

PPP/BKI:Ø +1660%

(120) On average, this cost group accounts for 3.2% of usage costs over the duration of the contract for public-private partnerships, 0.3% for KGSt and 0.2% for BKI (see below paragraph 155).

b) NOTES:

(121) Other PPP operating costs are reported for eight of the 16 PPP projects and are in most cases significantly higher than the KGSt and BKI key figures.

PPP: Costs of other cost groups

(122) This cost group was split in some PPP projects. This cost group was found to contain costs that could be allocated to other cost groups in DIN 18960 (e.g. overheads in cost group 200; insurance costs in cost group 370).

Additional project financing costs

(123) Projects with project financing are a special case group. In comparison to financing with forfeiting and waiver of defence, the project company incurs additional ongoing costs for auditing and tax advice, costs for guarantees and letters of intent as well as due diligence (e.g. monitoring compliance with service levels by external experts).

2.12 DIN 18960 CG 400 — Repairs

a) RESULTS:

DIN 18960 CG 400	PPP/KGST
Minimum	-27%
1st quartile	23%
Median (MED)	57%
3rd quartile	126%
Maximum	291%
Average value (AV)	74%
Average weighted value (AWV)	67%
Ø MED/AV/AWV	66%

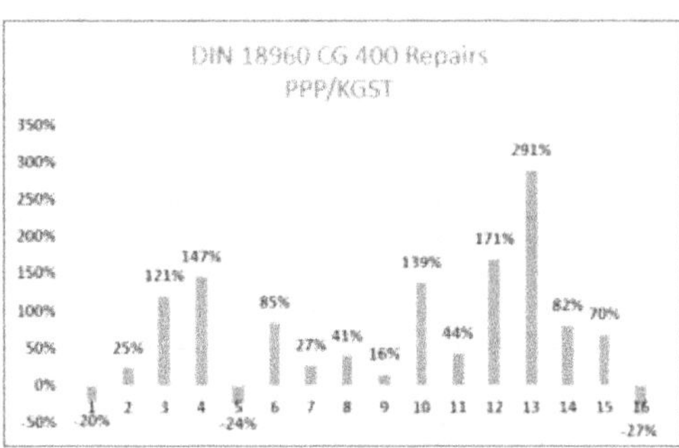

n=16

DIN 18960 CG 400	PPP/BKI
Minimum	-63%
1st quartile	-47%
Median (MED)	-30%
3rd quartile	-6%
Maximum	360%
Average value (AV)	10%
Average weighted value (AWV)	-9%
Ø MED/AV/AWV	-10%

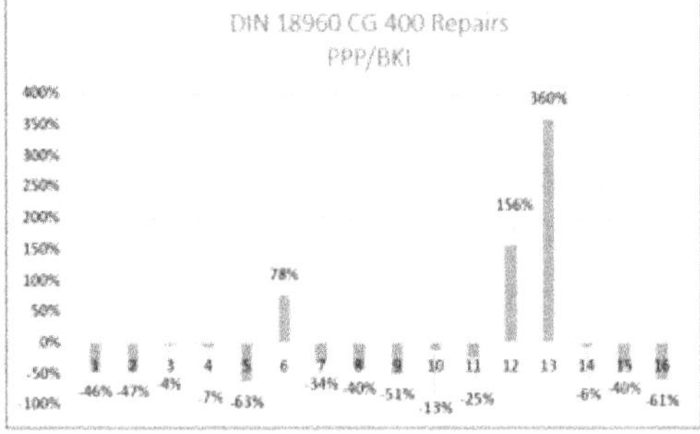

n=16

Table 2.19: Repair costs PPP/KGSt and PPP/BKI

(124) The PPP repair costs are on average of mean values 66% higher than the KGSt key figures. 13 of the 16 PPP budgets are over the KGSt value and three are under.

Repairs:
PPP/KGSt: Ø +66%

(125) The PPP repair costs are on average of mean values 10% lower than the BKI key figures. Three PPP projects are above the BKI key figure and 13 are below.

PPP/BKI:Ø -10%

(126) On average, repair costs account for 12.6% of usage costs over the duration of the contract for public-private partnerships, 7.4% for KGSt and 13.1% for BKI (see below paragraph 155).

b) NOTES:

(127) KGSt repair budgets are significantly lower than PPP and BKI budgets.

2.13 DIN 18960 CG 200 (pro rata), CG 350, CG 400 — Maintenance costs
2.13.1 Costs

a) RESULTS:

(128) The PPP maintenance costs are on average 137% higher than the KGSt values. All 16 PPP budgets are over the KGSt key figure.

(129) The PPP maintenance costs are on average 15% lower than the BKI key figures. Six PPP projects are above the BKI key figure and ten are below.

(130) On average, maintenance costs account for 20.2% of usage costs over the duration of the contract for public-private partnerships, 8.6% for KGSt and 21.5% for BKI (see below paragraph 155).

DIN 18960 CG 200+350+400	PPP/KGSt
Minimum	24%
1st quartile	105%
Median (MED)	137%
3rd quartile	199%
Maximum	311%
Average value (AV)	148%
Average weighted value (AWV)	126%
Ø MED/AV/AWV	137%

n=16

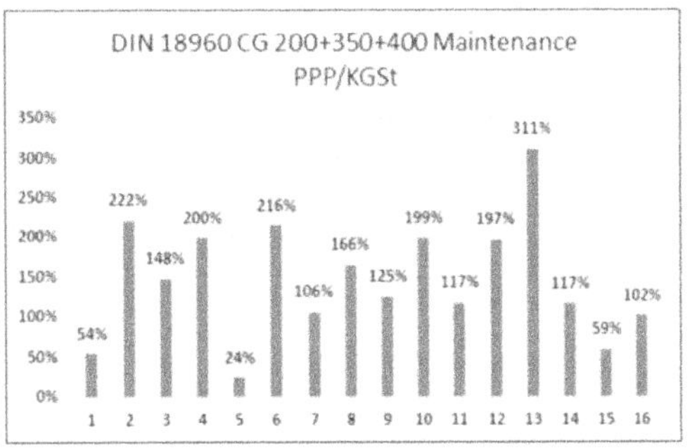

Maintenance:
PPP/KGSt: Ø +137%

DIN 18960 CG 200+350+400	PPP/BKI
Minimum	-56%
1st quartile	-34%
Median (MED)	-17%
3rd quartile	2%
Maximum	49%
Average value (AV)	-12%
Average weighted value (AWV)	-14%
Ø MED/AV/AWV	-15%

n=16

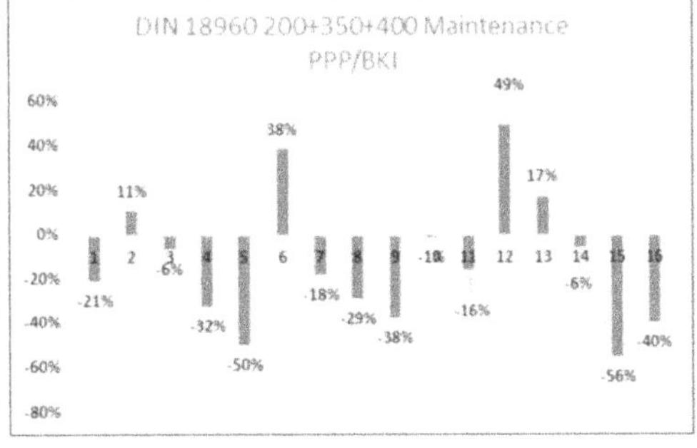

PPP/BKI:Ø -15%

Table 2.20: Maintenance costs PPP/KGSt and PPP/BKI

b) NOTES:

(131) Maintenance costs consist of the proportionate property management costs attributable to maintenance (DIN 18960 CG 200), maintenance and inspection costs (CG 350) and repair costs (CG 400).

Maintenance: pro rata property management + maintenance and inspection + repairs

(132) PPP maintenance budgets are significantly higher than KGSt and slightly lower than BKI.

KGSt actual significantly under PPP and BKI

2.13.2 Maintenance budget as a percentage of restoration costs

a) RESULTS:

(133) On average, PPP maintenance budgets account for 1.6% of restoration costs p.a. and are therefore significantly higher than KGSt (0.6%), markedly higher than the KGSt target budget (1.2%) and slightly lower than BKI (1.7% p.a.).

DIN 18960 CG 200+350+400 in % renovation costs p.a.	PPP
Minimum	1,2%
1st quartile	1,4%
Median (MED)	1,6%
3rd quartile	1,7%
Maximum	2,5%
Average value (AV)	1,6%
Average weighted value (AWV)	1,8%
Ø MED/AV/AWV	1,6%

n=16

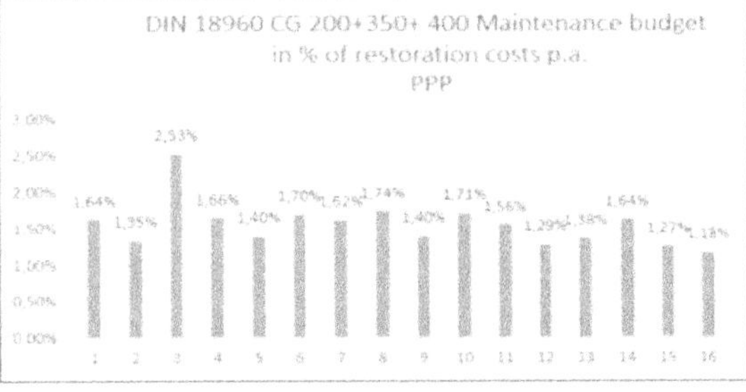

Maintenance budget in % of renovation costs p.a.:
PPP: 1.6%
KGSt actual: 0.6%,
KGSt target: 1.2%
BKI: 1.7%

DIN 18960 CG 200+350+400 in % renovation costs p.a.	KGST actual
Minimum	0,4%
1st quartile	0,6%
Median (MED)	0,6%
3rd quartile	0,7%
Maximum	0,8%
Average value (AV)	0,6%
Average weighted value (AWV)	0,6%
Ø MED/AV/AWV	0,6%

n=16

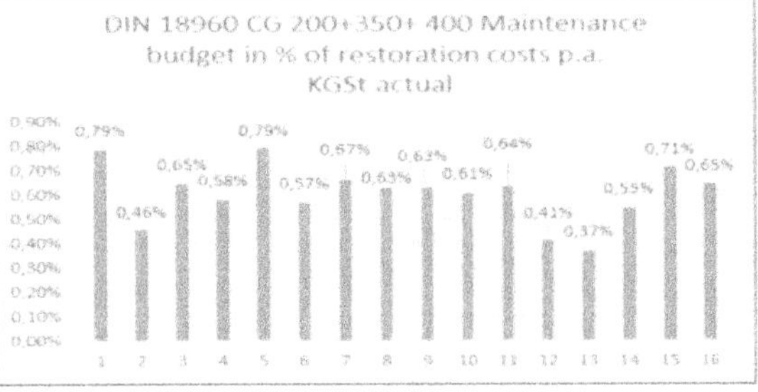

Result generally corresponds to the PPP school study (2019), PPP budgets are higher

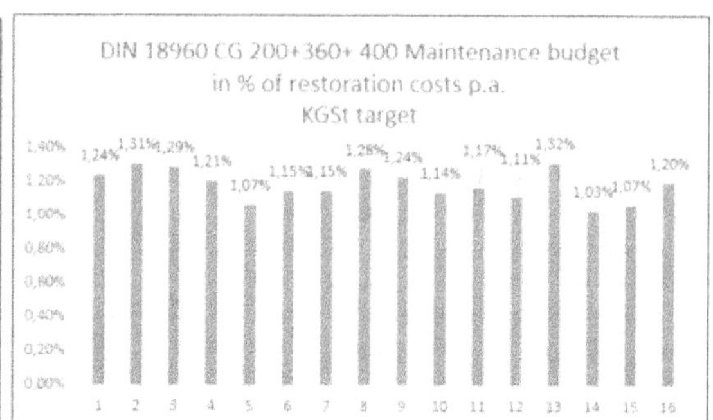

DIN 18960 CG 200+350+400 in % renovation costs p.a.	KGST target
Minimum	1,0%
1st quartile	1,1%
Median (MED)	1,2%
3rd quartile	1,3%
Maximum	1,3%
Average value (AV)	1,2%
Average weighted value (AWV)	1,2%
Ø MED/AV/AWV	1,2%

n=16

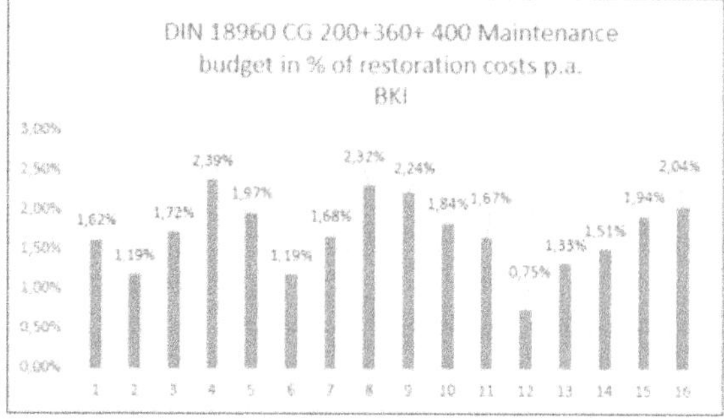

DIN 18960 CG 200+350+400	BKI
Minimum	0,7%
1st quartile	1,5%
Median (MED)	1,7%
3rd quartile	2,0%
Maximum	2,2%
Average value (AV)	1,7%
Average weighted value (AWV)	1,7%
Ø MED/AV/AWV	1,7%

n=16

**Table 2.21: Maintenance budget as a percentage of restoration costs
PPP/KGSt actual, BKI, KGSt target**

b) NOTES:

(134) The result is largely in line with the results of the PPP school study (2019, paragraph 108 ff.). The KGSt actual budgets are significantly below PPP, BKI and also the KGSt target budget recommendations. However, the average PPP maintenance budgets of the 16 PPP projects in this case group are significantly higher than the budgets of the 34 PPP new-build projects in the PPP school study (1.2% per annum).

(135) The differing maintenance budgets have implications for structural damage due to neglected maintenance and the expected residual value of the properties at the end of the contract (see below, paragraphs 153 and 162 f.).

Impact on structural damage and useful life/residual value

2.13.3 Qualitative aspects of PPP maintenance

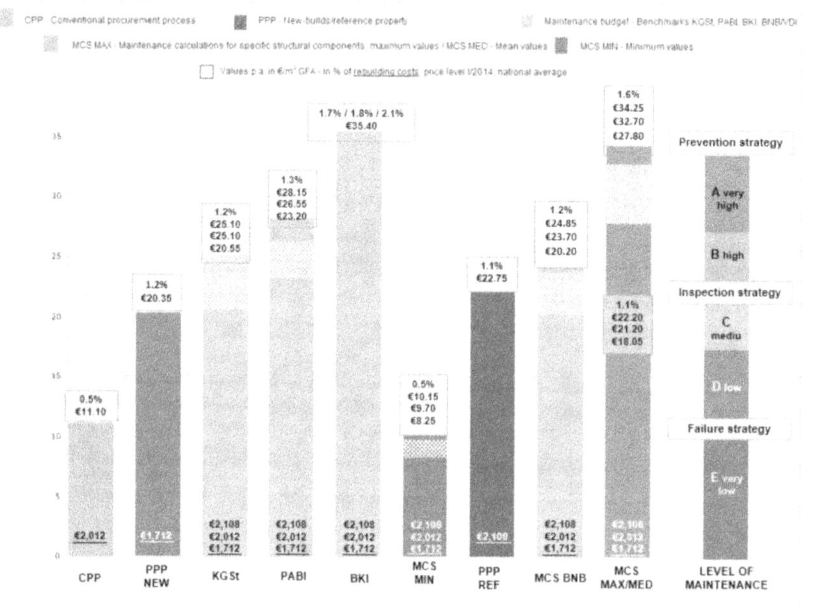

Table 2.22 Actual and target allocation of maintenance budgets over 25 years p.a. for 807 conventional schools and 34 PPP new-build school projects, pricing status I/2014, PPP school study (2019)

(136) In the PPP school study (2019), the analysis of part-specific maintenance calculations for a PPP project and the recommendations of the BNB guidelines (evaluation system for sustainable construction) for educational buildings showed that the KGSt target maintenance budget (1.2% of restoration costs per annum) facilitates a moderate maintenance level over 25 years and is essential for achieving the normal useful life of schools. With a budget of 0.6% per annum, on the other hand, only a very low level of maintenance (in terms of a contingency plan) can be achieved. In contrast, a PPP budget of 1.6% per annum enables a high level of maintenance.

(137) Furthermore, PPP maintenance management is marked by a number of key organisational factors which are crucial to ensuring the quality of the maintenance work. The incentive system established in the contract plays a central role here.

> Incentive structures in the PPP contract

(1) Cost ceilings for maintenance, service levels, bonus/malus rules

(138) In the PPP contract, it is typical for upper cost limits to be agreed for the maintenance fees. This works in tandem with agreeing on service levels, which specify the condition in which the key components of the property must be maintained over the entire term of the contract. Where targets are not met, response and correction rules apply, which determine the period within which defects must be rectified; if this is not done within the specified period, remuneration may be reduced (malus or bonus reductions).

> PPP costs risk from quality requirements

(139) A 2020 survey of the construction and finance departments of the 300 largest German cities, among others, found that users are not generally guaranteed fixed response and correction times and that, in conventional practice, there are no penalties for failing to rectify defects[14].

> Conventional process with no sanctions

(2) Part-specific maintenance calculations

(140) A PPP company that assumes this kind of contractual cost risk must take measures to minimise that risk. The PPP personnel report that no offer may be submitted without a part-specific maintenance calculation. This calculates how often and at what cost each part needs to be maintained, repaired or, if necessary, replaced over the term of the contract. These part-specific maintenance calculations require very specialised technical and financial expertise, which determines the calculation of the charges. Therefore, in the PPP process, quality defines the budget. It is therefore no coincidence that PPP maintenance budgets are in line with the relevant benchmarks, as the budget requirements are determined on a property-specific basis.

> Risk reduction with the most accurate and part-specific calculation possible

(141) In contrast, conventional maintenance budgets have generally been determined by general budgetary considerations; with budgets that are often far too low, only a very low level of maintenance can be guaranteed. The above-mentioned survey also revealed that in general there are no specifications for the preparation of part-specific maintenance calculations. It is worth noting that 79% of respondents were unable to specify how high their annual maintenance budgets are as a percentage of restoration costs[15]. There is evidence to suggest that this indicator in particular is a very effective tool for building administrations to raise budgets to an appropriate level based on operator risks (see PPP school study 2019, paragraph 210).

> Conventional process: Budgets with no precise maintenance calculation

(3) Reserve accounts with a share in any remaining credit balances

(142) Another typical feature of PPP contracts is the reserve account, which is used to deposit or record funds allocated to maintenance budgets that are not required, so that they are available for immediate use should the need arise. Any remaining balance at the end of the contract is usually divided in half.

> Reserve account facilitates immediate access to funds

(143) In the conventional model, reserve accounts such as these are uncommon. The administration describes these as a "quantum leap," because in conventional practice a complex budget allocation process has to be carried out for each individual repair measure, which often results in the repair being postponed due to conflicting priorities with other local authority responsibilities.

> This does not exist in the conventional process. PPP reserve accounts are a "quantum leap"

This is one of the reasons for the backlog in local authority investment identified by the KfW local authorities panel 2014, for example in schools, in the amount of €54.76 billion[16].

(4) Sufficient number of personnel

(144) The cost analysis shows that personnel costs account for a significantly higher proportion of usage costs in public-private partnerships than under the conventional system. Part of the maintenance calculation is also the need to maintain sufficient numbers of personnel in order to manage the contractual risks and successfully fulfil the contract. This requires, for example, caretakers who are able to carry out small repairs themselves, or the graduated organisation of the project with well-trained staff on site and in the head office. Naturally, remuneration is based on performance-related payment elements.

Sufficient number of well-trained staff

(145) Conventionally, in contrast, there is a considerable shortage of staff[17], and there are often not enough staff to spend the too-low maintenance budget. In the event of doubt, this will lead to further cuts in the "clearly not required" budget the following year.

Severe staff shortages in conventional process

(146) As a result, it is clear that PPP maintenance management differs from the conventional process in a number of key organisational aspects and that the incentive system embedded in the PPP contract is a major reason why it is possible to achieve a sustainable and high-quality level of maintenance.

2.14   Usage costs

2.14.1 Usage costs excluding risk costs

Usage costs excluding risk costs	PPP/KGSt
Maximum	-10%
1st quartile	-9%
Median (MED)	0%
3rd quartile	4%
Minimum	20%
Average value (AV)	0%
Average weighted value (AWV)	-3%
Ø MED/AV/AWV	-1%

n=16

Usage costs excluding risk costs	PPP/BKI
Maximum	-28%
1st quartile	-21%
Median (MED)	-13%
3rd quartile	-9%
Minimum	-7%
Average value (AV)	-15%
Average weighted value (AWV)	-17%
Ø MED/AV/AWV	-15%

n=16

Table 2.23: Usage costs excluding risk costs PPP/KGSt and PPP/BKI

a) RESULTS:

(147) The PPP usage costs excluding risk costs are on average 1% lower than the KGSt figures. Eight PPP projects undershot the KGSt key value (between -2% and -11%), eight exceeded it (between +1% and +20%), five of them by a small margin of between 1% and 4%).

Usage costs excluding risk costs: PPP/KGSt: Ø -1%

(148) The PPP usage costs excluding risk costs are on average 15% lower than the BKI key figures. All 16 PPP projects came in under the BKI key figure.

PPP/BKI: Ø -15%

b) NOTES:

(149) Despite higher costs for maintenance, property management, cleaning and other operating costs, the PPP usage costs are on average slightly below the KGSt costs, even without a risk assessment. This is made possible by lower capital costs (due to lower construction costs) and lower energy consumption costs. In five of the seven projects with higher costs, the additional costs are relatively low (+0.8% to +4.0%).

(150) The BKI key figures are relatively clearly undercut. This is made possible by efficient construction processes with low construction costs and short construction times as well as savings in energy management.

2.14.2 Usage costs including risk costs

Usage costs including risk costs	PPP/KGSt
Maximum	-12%
1st quartile	-9%
Median (MED)	-2%
3rd quartile	3%
Minimum	17%
Average value (AV)	-2%
Average weighted value (AWV)	-4%
Ø MED/AV/AWV	-3%

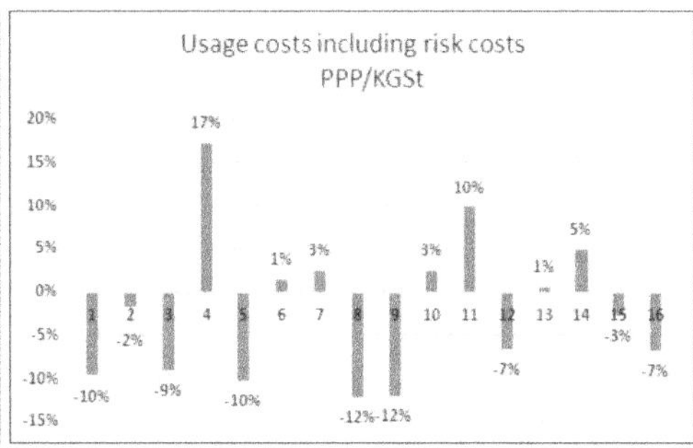

n=16

Usage costs including risk costs	PPP/BKI
Maximum	-28%
1st quartile	-20%
Median (MED)	-13%
3rd quartile	-9%
Minimum	-7%
Average value (AV)	-15%
Average weighted value (AWV)	-16%
Ø MED/AV/AWV	-15%

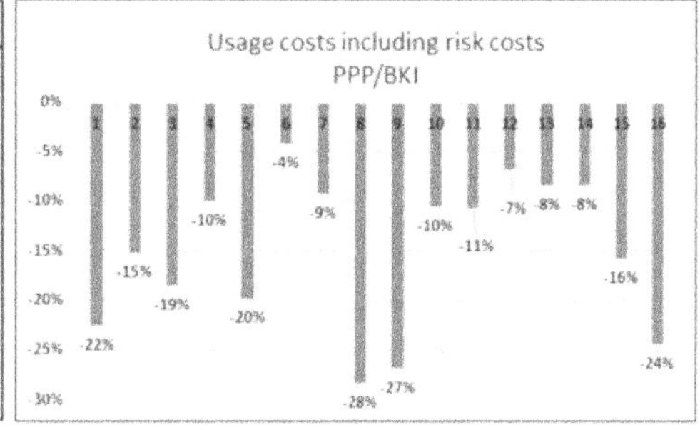

n=16

Table 2.24: Usage costs including risk costs PPP/KGSt and PPP/BKI

a) RESULTS:

(151) The PPP usage costs including risk costs are 3% lower than the KGSt figures.

(152) The PPP usage costs including risk costs are on average 15% lower than the BKI key figures.

Usage costs including risk costs:
PPP/KGSt: Ø -3%
PPP/BKI: Ø -15%

b) NOTES:

(153) The difference between this and results for usage costs excluding risk costs is primarily due to the inclusion of risk costs for structural damage caused by low maintenance budgets under the KGSt model (amount: 2% of production costs, indexed to the end of the contract). This estimate is likely to be at the lower end. For instance, there is the practical example of the city of Neuwied where neglected maintenance caused structural damage to 16 schools amounting to 7% of the restoration costs (see PPP school study (2019), paragraph 139).

Building damage as a result of neglected maintenance

(154) Theoretically, with public-private partnerships other costs could also be allocated to risk costs (such as higher interest rates, guarantee costs and due diligence costs; see Projektdokumentation PPP-Berufskolleg Duisburg, p. 40).

2.14.3 Share of each cost group in the usage costs

(155) If the distribution of usage costs across the individual cost groups of the three alternatives is considered, the following picture emerges:

(156) Capital costs are the largest single cost group for all three models. The

second largest cost item for public-private partnerships and BKI is repairs (third largest for KGSt after cleaning costs). With public-private partnerships, this is followed by property management costs (if portions of cost group 390 are included), cleaning, maintenance and inspection, and utilities.

DIN 18960	Cost groups	PPP ø		KGSt ø		BKI ø		
CG 100	Capital costs	59,2%	1	67,7%	1	59,0%	1	Share of each cost group in the usage costs
CG 100	Interest final financing	21,4%		23,4%		20,4%		
CG 100	Interest final financing							
CG 200	Property management	6,7%	4 (3)	6,0%	5	4,6%	6	
CG 210	Personnel costs	5,9%						
CG 210	City personnel costs	0,8%						
CG 300	Operating costs	21,4%		17,5%		23,2%		
CG 310	Supply**	5,2%	6	6,4%	4	6,5%	5	
CG 311	Water	0,3%		0,4%		0,5%		
CG 312	Heating	2,3%		3,0%		3,3%		
CG 313	Electricity	2,5%		2,4%		2,5%		Different budget allocation
CG 320	Waste disposal	1,0%	8	1,0%	6	0,8%	8	
CG 330	Cleaning	8,5%	3 (4)	8,2%	2	6,8%	4	
CG 350	Maintenance and inspection	6,1%	5	0,7%	7	7,4%	3	
CG 350	Energy management	0,5%				2,1%		
CG 370	Taxes, contributions, insurance	0,9%	9	0,4%	8	1,1%	7	
CG 391	Canteen	0,7%		0,7%		0,6%		
CG 393	Other operating costs	3,2%	7	0,3%	9	0,2%	9	
CG 400	Repairs	12,6%	2	7,4%	3	13,1%	2	
CG 460	Risk of construction damage (maintenance)*			1,6%				
CG 200/350/400	Maintenance budget per year	20,2%		8,6%		21,5%		
CG 100-400	Usage costs	100%		100%		100%		
CG 200-400	Usage costs excluding capital costs	41%		31%		41%		
CG 300+400	Maintenance costs and repairs	34%		25%		36%		

Table 2.25: Allocation of usage costs for PPP, KGSt and BKI

(157) BKI maintenance and inspection costs (third largest item) are higher than the costs of cleaning, utilities and property management.

(158) The costs for KGSt property management are in fourth place behind the costs of utilities. The costs of maintenance and inspection account for a smaller fraction than waste disposal costs, which rank only eighth for public-private partnerships and BKI.

(159) The biggest differences between KGSt on the one hand and public-private partnerships and BKI on the other are in maintenance costs, with KGSt values over 50% lower than for the other two. In the case of public-private partnerships, property management costs are a priority and utility costs benefit from good energy management.

2.15   Comparison of residual value changes

DIN 18960 CG 500 Residual value	PPP/KGST
Minimum	14%
1st quartile	21%
Median (MED)	28%
3rd quartile	33%
Maximum	47%
Average value (AV)	29%
Average weighted value (AWV)	25%
ø MED/AV/AWV	27%

n=16

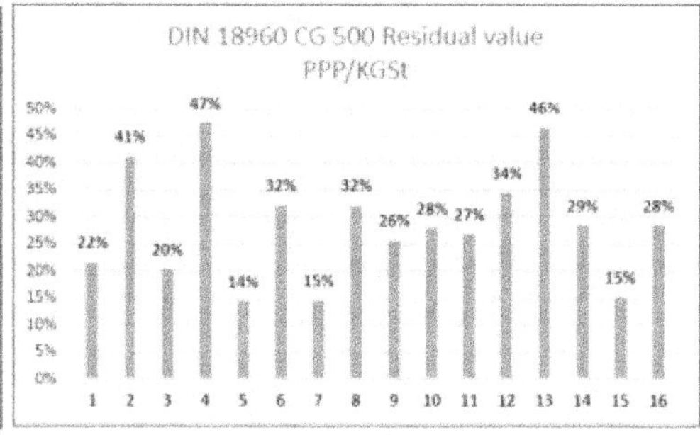

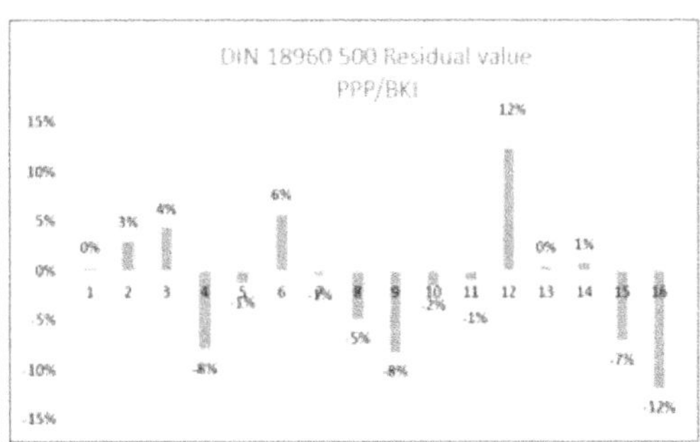

DIN 18960 CG 500 Residual value	PPP/BKI
Minimum	-12%
1st quartile	-5%
Median (MED)	-1%
3rd quartile	1%
Maximum	12%
Average value (AV)	-1%
Average weighted value (AWV)	0%
Ø MED/AV/AWV	0%

n=16

Table 2.26: Comparison of value development

a) RESULTS:

(160) The projected PPP residual value at the end of the contract term is an average of 27% higher than the KGSt alternative and on a par with the BKI alternative.

Residual value:
PPP/KGSt: Ø +27%
PPP/BKI: Ø 0%

b) NOTES:

(161) When calculating the projected residual value, three distinct aspects must be considered: (1) interdependence of maintenance budget and residual value/useful life, (2) allowance for hidden reserves and (3) price changes.

(162) On (1): The size of the maintenance budget and the associated various maintenance strategies are crucial factors in determining the residual value of a property as they have a significant influence on the potential risks and rewards of the changes in residual value. It seems to be certain that a connection exists. What is not certain is its size. This will require extensive empirical evidence. The PPP school study (2019) presents initial findings based on 800 conventional schools, 50 PPP projects and reports from local authorities on the effects of low maintenance budgets on structural damage, useful life and residual values (PPP-Schulstudie 2019, paragraph 137 ff.).

Interdependence of residual value and maintenance budget

(163) Example: Achieving the standard useful life of 80 years for school buildings and ensuring a moderate level of maintenance throughout the operating phase requires a maintenance budget of 1.1% of the restoration costs per year in the life cycle phase of the first 25 years; the residual value after 25 years would then be €16.9 million. Should the actual available maintenance budget be lower (0.5% per annum), the useful life would be reduced by 30% to 56 years and the residual value would fall to €13.1 million; should the maintenance budget be higher (1.9% per annum), the useful life would be extended to 91 years and the residual value would increase to €18.8 million. The 30% reduction in useful life with very low maintenance budgets can be empirically substantiated (see PPP-Schulstudie 2019, loc. cit.). As far as is apparent, there is no empirical evidence to date on extending the useful life with higher maintenance budgets.

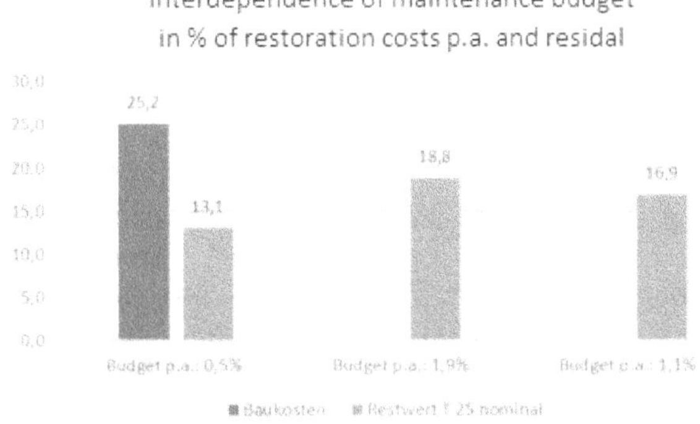

Table 2.27: Interdependence of maintenance budget and residual value

These estimates are based on the assumption that the rate at which the useful life is extended as the maintenance budget increases is somewhat slower than the rate at which it is shortened as the budget decreases.

(164) On (2): Generally, the residual value is determined by amortising the production costs/construction costs. If, for example, the PPP construction costs of €17.7 million are 30% below the BKI construction costs of €25.2 million with a comparable construction standard, the PPP residual value with straight-line amortisation after 25 years would be €12.1 million, €5.2 million below the BKI residual value of €17.3 million. That is implausible. This shows that the efficient PPP construction process has created hidden reserves to the benefit of the local authority when compared to the average conventional costs. The projected PPP residual value of the property at the end of 25 years must therefore be calculated with the hidden reserves included.

Residual value and hidden reserves

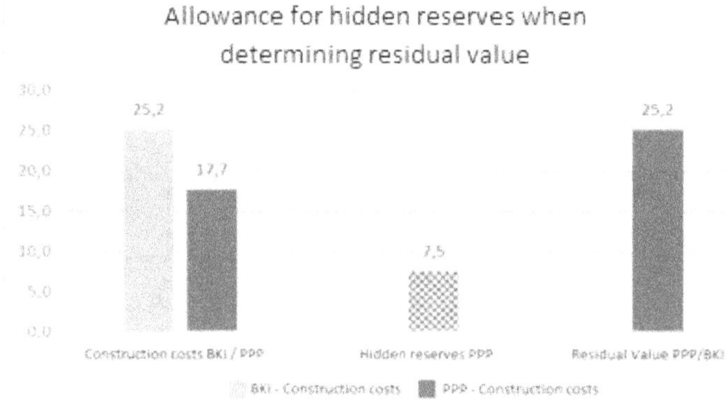

Table 2.28: Allowance for hidden reserves when determining residual value

(165) On (3): Economic viability studies should make the expected cost trend transparent (see Hesse State Audit Office, audit report dated 26 March 2015 on the Offenbach district PPP school project). This is why the operating costs are indexed. Consequently, the projected price changes must also be taken into account when determining the residual value.

Residual value and indexing

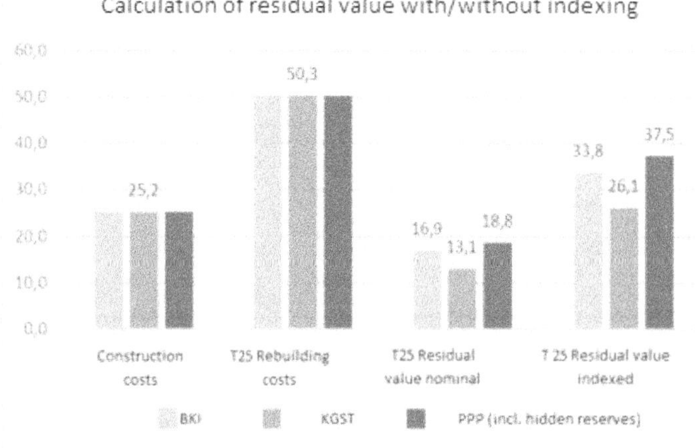

Table 2.29: Calculation of residual value with/without indexing

(166) In the example, the restoration costs with an annual price increase of 2.92% (DESTATIS, construction price index, non-residential/commercial buildings between 2007 and 2021) amount to €50.3 million over 25 years. Depending on the maintenance strategy, the (indexed) residual value is expected to be between €26.1 million and €37.5 million.

2.16   Life cycle costs

2.16.1 Life cycle costs excluding risk costs

a) RESULTS:

Life cycle costs excluding risk costs	PPP/KGSt
Maximum	-28%
1st quartile	-22%
Median (MED)	1%
3rd quartile	10%
Minimum	56%
Average value (AV)	-1%
Average weighted value (AWV)	-9%
Ø MED/AV/AWV	-3%

n=16

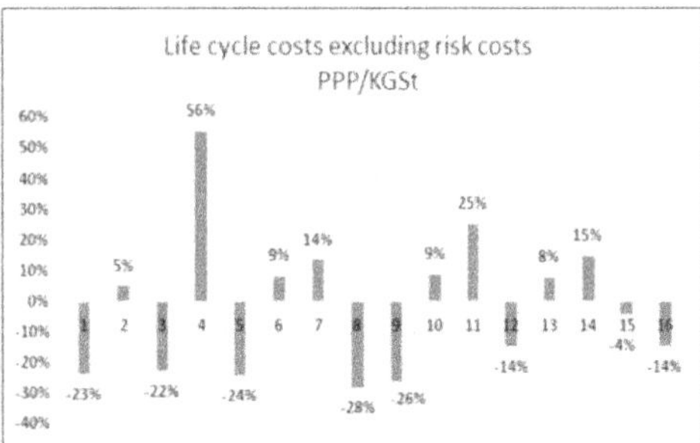

Life cycle costs excluding risk costs	PPP/BKI
Maximum	-56%
1st quartile	-48%
Median (MED)	-29%
3rd quartile	-20%
Minimum	-10%
Average value (AV)	-33%
Average weighted value (AWV)	-35%
Ø MED/AV/AWV	-32%

n=16

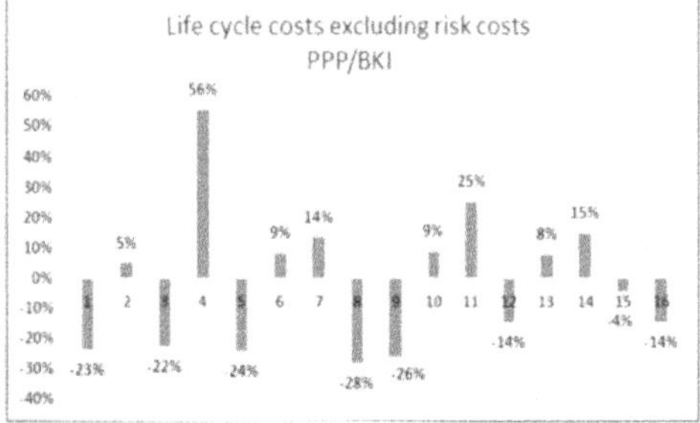

Table 2.30: Life cycle costs excluding risk costs PPP/KGSt and PPP/BKI

(167) The PPP life cycle costs excluding risk costs are on average 3% lower than the KGSt model. Eight PPP projects are below the KGSt model (between -4% and -28%) and eight are higher (between +5% and +56%).

Life cycle costs excluding risk costs: PPP/KGSt: Ø -3%

(168) The PPP life cycle costs excluding risk costs are on average 32% lower than the BKI model. All 16 PPP projects came in under the BKI model (between -9% and -56%).

PPP/BKI: Ø -32%

b) NOTES:

(169) Life cycle costs are defined as usage costs minus residual value. This includes resource consumption during the operating phase.

(170) The life cycle costs excluding risk costs are calculated without quantifying the maintenance risks. This means that no structural damage is included in the usage costs for the KGSt model. Furthermore, the residual value is determined in the same way across all models.

(171) The percentage comparison results are higher than the results for usage costs because the assessment basis for the percentage comparison (usage costs minus residual value) is lower.

2.16.2   Life cycle costs including risk costs [18]

a) RESULTS:

(172) The PPP life cycle costs including risk costs are on average 35% lower than the KGSt model. All 16 PPP projects came in under the KGSt model (between -2% and -64%).

Life cycle costs including risk costs: PPP/KGSt: Ø -35%

(173) The PPP life cycle costs including risk costs are on average 34% lower than the BKI model. All PPP projects are below the BKI model (between -13% and -58%).

PPP/BKI: Ø -34%

Life cycle costs including risk costs	PPP/KGSt
Maximum	-64%
1st quartile	-44%
Median (MED)	-37%
3rd quartile	-20%
Minimum	-2%
Average value (AV)	-33%
Average weighted value (AWV)	-34%
Ø MED/AV/AWV	-35%

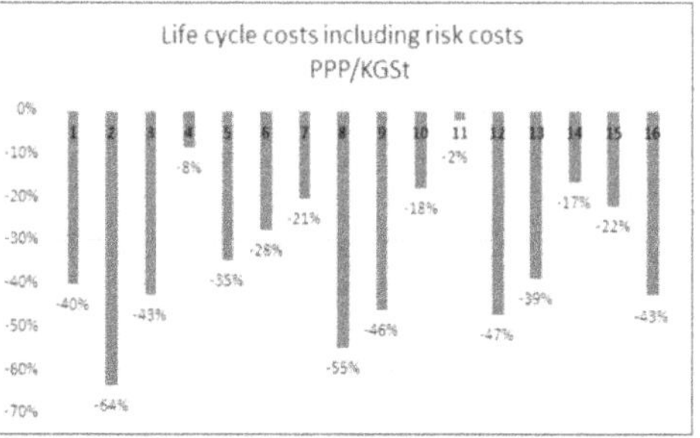

n=16

Life cycle costs including risk costs	PPP/BKI
Maximum	-58%
1st quartile	-47%
Median (MED)	-31%
3rd quartile	-19%
Minimum	-13%
Average value (AV)	-34%
Average weighted value (AWV)	-37%
Ø MED/AV/AWV	-34%

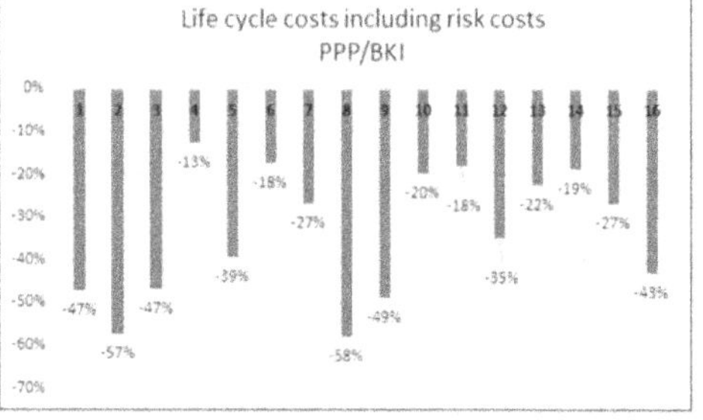

n=16

Table 2.31: Life cycle costs including risk costs PPP/KGSt and PPP/BKI

b) NOTES:

(174) The comparison between public-private partnerships and KGSt shows how significant the maintenance budget is for the residual value. Maintenance budgets more than 50% below the recommendations of relevant benchmarks result in structural damage[19] and significantly reduced useful lives and residual values[20]. In this regard, it seems highly plausible that insufficient maintenance budgets are a primary reason for the €54.67 billion investment backlog in schools (KfW-Kommunalpanel 2024).

Impact of low maintenance budgets on useful life and residual value is significant

(175) The comparison between public-private partnerships and BKI changes only slightly when considering the risk costs.

(176) It should be noted that indexing significantly affects the overall result (see below).

2.16.3   Life cycle costs including risk costs and additional VAT income

a) RESULTS:

(177) When including the additional VAT income generated by public-private partnerships, the PPP advantage in life cycle costs including risk costs increases to 37% compared to the KGSt model and to an average of 36% compared to the BKI model.

Life cycle costs including risk costs and additional VAT income:
PPP/KGSt: Ø -37%
PPP/BKI: Ø -36%

Life cycle costs including risk costs+VAT	PPP/KGSt
Maximum	-66%
1st quartile	-45%
Median (MED)	-39%
3rd quartile	-22%
Minimum	-5%
Average value (AV)	-35%
Average weighted value (AWV)	-36%
Ø MED/AV/AWV	-37%

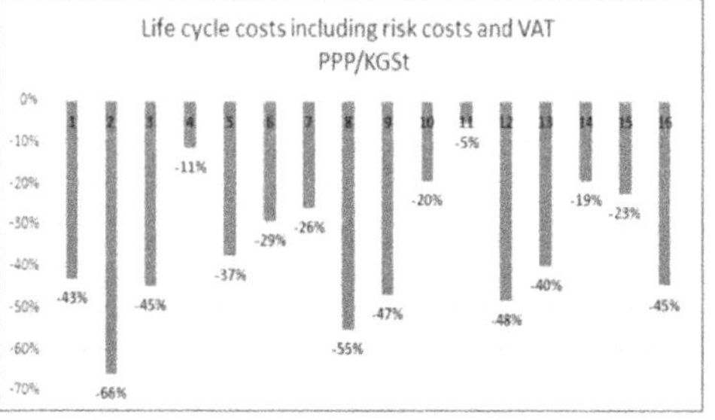

n=16

Life cycle costs including risk costs+VAT	PPP/BKI	
Maximum	-60%	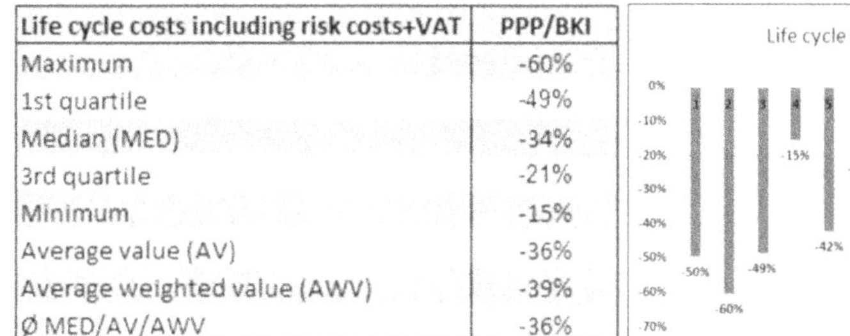
1st quartile	-49%	
Median (MED)	-34%	
3rd quartile	-21%	
Minimum	-15%	
Average value (AV)	-36%	
Average weighted value (AWV)	-39%	
Ø MED/AV/AWV	-36%	

n=16

Table 2.32: Life cycle costs including risk costs and additional VAT income

b) NOTES:

(178) PPP personnel costs are subject to VAT, while the costs of public employees are exempt from VAT; public-private partnerships therefore lead to additional VAT income for federal, state and local authorities.

(179) The additional income comes to approximately 2% of the life cycle costs.

(180) Note: Unlike in several EU Member States, there is no tax refund system in Germany through which PPP local authorities could have their VAT refunded by the tax office. This is designed to eliminate the competitive disadvantages of realising VAT under private law.

In Germany, no VAT tax refund

2.16.4 Results overview of PPP new-build projects

European Comparative PPP Study - Germany			GREENFIELD PROJECTS Projects 1-16	
			PPP/KGST	PPP/BKI
DIN 276	CG 200-700	Construction costs	-17% (-15% / - 20%)	
		Cost overruns (n=15)	0,6% (Median 0%)	
DIN 277		Construction time	-30%	
		Time overruns (n=15)	2,3% (Median 0%)	
DIN 18960	CG 100	Capital costs	-15%	-15%
	CG 200	Object management costs	12%	23%
	CG 300	Operating costs	19%	-24%
	CG 310	Water, heating, electricity	-16%	-33%
	CG 320	Disposal costs	0%	0%
	CG 330/40	Cleaning costs	5%	9%
	CG 350	Servicing&inspection	685%	-21%
	CG 360	Energy management	0%	-2%
	CG 370	Taxes, contributions, assurance	179%	-7%
	CG 390	Other operating costs*	455%	1660%
	CG 400	Repair costs	66%	-10%
	CG 200/350/400	Maintenance budget	137%	-15%
		in % p.a. PPP / KGST-TARGET	1,6%	1,2%
		in % p.a. KGST-ACT / BKI	0,6%	1,7%
Usage costs without risk costs			-1%	-15%
Usage costs including risk costs			-3%	-15%
Residual value			27%	0%
Life cycle costs without risk costs			-3%	-32%
Life cycle costs including risk costs			-35%	-34%
Life cycle costs including risk costs and VAT			-37%	-36%

* Percentage of the cost group CG 390 in the usage costs: PPP 3.2% (of which approx. 3% costs are attributable to KG 200), KGST: 0.3%, BKI: 0.2%

Table 2.33: Overview of results of PPP new-build projects

(181) Compared to the KGSt model, the pure PPP usage costs over 25 years are 1% or, including risk costs, 3% below the costs of the KGSt model. If the effect

PPP/KGSt excluding/including Risk costs:

of the various maintenance strategies on structural damage, useful lives and residual values is taken into account, the PPP advantage increases to 34%, and to 36% including additional VAT income.

Usage costs -1% / -3% Life cycle costs: -3% / -35% / including additional VAT income: -37%

(182) A comparison of the PPP/KGSt models shows in particular:

- PPP capital costs (-15%) and energy supply costs (-16%) are lower, while the PPP costs for property management (+12%), cleaning (+5%), maintenance and inspection (+685%), insurance (+179%), other operating costs (+455%) and repairs (+66%) are higher than the comparable KGSt costs. On balance, the PPP construction cost advantage (15-20%) results in a pure cost advantage of 1% in usage costs.

PPP with partly higher operating costs

- It should be noted that the additional costs are counterbalanced by higher levels of quality: For example, the second largest maintenance cost item is of above-average quality, while the reality of local authority procurement illustrated in the KGSt model means that only very low levels of maintenance can be carried out due to personnel and budget restrictions.

Additional costs are partially compensated for by significantly higher quality, especially maintenance

- In the pure cost analysis, the financing costs include additional costs due to higher interest rates applicable to public-private partnerships, although these additional costs are offset by qualitative benefits (interest rate hedging for the fixed construction cost price and additional risk hedging in project financing due to quality controls by the refinancing bank).

Financing

- With regard to cleaning costs, one example shows that the PPP disadvantage in the higher cleaning costs per m^2 of gross floor area can translate into a PPP cost advantage per m^2 of annual cleaning area due to longer cleaning intervals.

Example cleaning

- The additional costs for property management (including other operating costs) are lower than the additional VAT charges included in the costs, i.e. the resulting costs for the PPP local authority are offset by additional tax income at the federal, state and local level. In other EU countries besides Germany, PPP local authorities can have these additional costs reimbursed through a tax refund system.

Property management additional VAT income for the federal and state authorities

- The very high percentage of PPP additional costs for cost group 370 and cost group 390 are related to the comparatively low conventional nominal costs, which account for only 0.4% and 0.3% of the KGSt usage costs, respectively.

Cost groups 370/390 low share of usage costs

- In this respect, it seems remarkable that the PPP usage costs are still (slightly) lower on average than the usage costs of the KGSt model, despite significantly higher quality in some cases. This is generally achieved through highly efficient construction processes and savings in energy consumption.

PPP efficiency in the construction process allows for lower costs and significantly higher quality

- The economic advantage of the PPP approach over the KGSt model becomes clear when evaluating the different maintenance strategies: Very low maintenance budgets carry the risk of structural damage, shortened useful lives and reduced residual values, while above-average budgets have a positive effect on useful lives and residual values. The quantitative effects can be determined from initial empirical studies (PPP-Schulstudie 2019). For the 16 PPP projects in question, the PPP advantage in terms of life cycle costs increased to 34% due to the evaluation of structural damage and differences in useful life and changes in residual value. Reports on investment backlogs of €54.76 billion in schools (KfW-Kommunalpanel 2024[21]), caused in part by a lack of maintenance, show that such orders of magnitude can be realistic.

PPP residual value advantage 34% when evaluating the differing levels of maintenance quality

(183) Compared to the BKI model, the pure PPP usage costs over a 25-year period are 15% lower than the usage costs of the BKI model. If the future residual value is included, the PPP advantage in terms of life cycle costs increases to 34%, and to 36% if the additional VAT income is included.

PPP/BKI: PPP advantage 15%-36%

(184) A comparison of the PPP/BKI models shows in particular:

- The PPP costs for debt servicing (-15%), energy supply (-32%), maintenance and inspection (-21%), insurance (-7%) and repairs (-10%) are lower than the comparable BKI values, while the costs for property management (+23%), cleaning (+9%) and other operating costs (+1660%) are higher. On balance, the PPP construction cost advantage (15-20%) results in a pure cost advantage of 15% in usage costs. **PPP operating costs mainly below BKI**

- The additional costs for property management (23%) are higher than the VAT.

- PPP costs are 21% lower than the BKI estimates for maintenance and inspection and 10% lower with regard to repairs, i.e. the BKI estimates are even higher than the costs of the KGSt model than the PPP costs. This means that conventional projects with these low maintenance budgets are significantly underfunded, even when evaluated according to the key performance indicator system of the German Chambers of Architects. **BKI confirms deficits from insufficient conventional maintenance budgets**

- The PPP advantage in terms of usage costs (15%) increases to 34% in terms of life cycle costs when the projected residual value is included. This is due to the fact that the higher BKI usage costs are calculated on a lower assessment basis (usage costs minus residual value) when determining the life cycle costs. **If future residual value is included in the life cycle costs, the PPP advantage is 34%**

2.16.5 Impact of indexing on the overall result

(185) The indexing level has a significant impact on the overall result. This applies in particular to the construction price index, which is relevant for indexing repair costs and the projected residual value. Accordingly, sensitivity analyses with an annual construction price increase rate of 2.0% and 1.5% were carried out for all projects. **Indexing is relevant**

Indexing residual value+repair costs with index 2.92% p.a.*	PPP/KGSt	PPP/BKI	
Usage costs excluding risk costs	-1%	-15%	**Increase in construction prices (2007-2021) 2.92% p.a.**
Usage costs including risk costs	-3%	-15%	
Life cycle costs excluding risk costs	-3%	-32%	
Life cycle costs including risk costs	-35%	-34%	
Life cycle costs including risk costs and VAT	-37%	-36%	
Sensitivity analysis — Indexing residual value and maintenance with 2% p.a.			**Sensitivity analysis: Construction price index with 2% p.a.**
Usage costs excluding risk costs	-2%	-15%	
Usage costs including risk costs	-3%	-15%	
Life cycle costs excluding risk costs	-4%	-27%	
Life cycle costs including risk costs	-26%	-28%	
Life cycle costs including risk costs and VAT	-28%	-30%	
Sensitivity analysis — Indexing residual value and maintenance with 1.5% p.a.			**Sensitivity analysis: Construction price index with 1.5% p.a.**
Usage costs excluding risk costs	-3%	-15%	
Usage costs including risk costs	-4%	-15%	
Life cycle costs excluding risk costs	-5%	-26%	
Life cycle costs including risk costs	-23%	-26%	
Life cycle costs including risk costs and VAT	-25%	-28%	

Table 2.34: Impact of construction price indexing on the overall result

(186) As a result, when indexed at 2% per annum, the PPP advantage in terms of life cycle costs including risk costs is reduced from 35% to 26% compared to the KGSt model and from 34% to 28% compared to the BKI model.

(187) With an annual increase in construction prices of 1.5%, the PPP advantage is reduced to 23% (KGSt) and 26% (BKI).

2.16.6 Partial results for schools, administrative buildings and project financing

European comparative PPP study (Germany)	SCHOOLS		ADMINISTRATION		PROJECT FINANCING	
	ØPPP-KGSt	ØPPP-BKI	ØPPP-KGSt	ØPPP-BKI	ØPPP-KGSt	ØPPP-BKI
	n=14		n=2		n=4	
DIN 276 CG 200-700 Construction costs PPP/CPP including CG 760	-16%		-20%		-20%	
DIN 277 Construction time	-30%		-31%		-27%	
DIN 276 CG 100 Capital costs	-13%	-13%	-19%	-19%	-15%	-15%
KG 200 Property management	12%	26%	7%	0%	-18%	-2%
KG 300 Operating costs	18%	-25%	45%	10%	22%	-14%
KG 310 Supply	-20%	-32%	-3%	-11%	15%	0%
KG 320 Waste management	0%	0%	0%	0%	0%	0%
KG 330/40 Cleaning	6%	5%	15%	66%	-17%	-17%
KG 350 Maintenance and Inspection	545%	-12%	296%	-17%	442%	-20%
KG 360 Energy management	0%	-7%	0%	0%	0%	0%
KG 370 Insurences, taxes	50%	-12%	279%	141%	174%	-24%
KG 390 Other operating costs	403%	746%	285%	7934%	382%	9%
KG 400 Repairs	67%	-6%	128%	117%	76%	-30%
KG 200/350/400 Maintenance budget	139%	-20%	206%	44%	107%	-28%
in % of renovation costs p.a. PPP/KGSt target	1,6%	1,2%	1,5%	1,1%	1,7%	1,1%
in % of renovation costs p.a. KGSt actual/BKI	0,6%	1,8%	0,5%	1,0%	0,7%	1,9%
Usage costs excluding risk costs	0%	-16%	-1%	-5%	-4%	-16%
Usage costs including risk costs	-2%	-16%	-3%	-5%	-5%	-16%
Residual Value	28%	-2%	33%	9%	20%	-1%
Life cycle costs excluding risk costs	-1%	-35%	-3%	-15%	-10%	-33%
Life cycle costs including risk costs	-32%	-35%	-38%	-26%	-29%	-33%
Life cycle costs including risk costs and VAT	-34%	-37%	-39%	-28%	-31%	-35%

Sensitivity analysis — Indexing residual value+repairs 2% p.a.						
Usage costs excluding risk costs	-1%	-17%	-2%	-7%	-5%	-17%
Usage costs including risk costs	-2%	-17%	-3%	-7%	-6%	-17%
Life cycle costs excluding risk costs	-2%	-30%	-5%	-14%	-10%	-29%
Life cycle costs including risk costs	-24%	-29%	-28%	-21%	-24%	-29%
Life cycle costs including risk costs and VAT	-25%	-31%	-30%	-23%	-25%	-30%
Sensitivity analysis — Indexing residual value+repairs 1.5% p.a.						
Usage costs excluding risk costs	-2%	-17%	-3%	-7%	-5%	-17%
Usage costs including risk costs	-3%	-17%	-4%	-7%	-6%	-17%
Life cycle costs excluding risk costs	-4%	-28%	-5%	-13%	-10%	-28%
Life cycle costs including risk costs	-21%	-27%	-25%	-20%	-22%	-28%
Life cycle costs including risk costs and VAT	-22%	-29%	-26%	-21%	-23%	-29%

*Arithmetic average

Table 2.35: Partial results for schools, administrative buildings and project financing

(188) A comparison of the PPP school and administrative buildings reveals in particular that under the KGSt model, the maintenance budgets for the administrative buildings (0.5% of restoration costs per annum) are even lower than those for the school buildings (KGSt: 0.6%; BKI). It is also worth noting that the maintenance cost key figures for the BKI model are almost 50% lower for administrative buildings (1%) than for schools (1.8%). This leads to a shorter useful life and lower residual values. Accordingly, the PPP results for administrative buildings are better in relative terms than the results for school buildings.

Partial results of PPP school buildings and PPP administrative buildings

(189) When comparing PPP projects with project financing, the construction cost advantage (20%) falls to 15% in terms of capital costs; the higher financing interest rates are evident here. Nevertheless, the result for the life cycle costs of project financing projects is only slightly lower than the overall result for PPP new-build projects.

Partial result of PPP project financing

3 Results of PPP renovation projects

European comparative PPP study — Germany		RENOVATION 100% Projekte 17-18	
		PPP/KGST	PPP/BKI
DIN 276 CG 200-700	Construction costs	-11%	
	Cost overruns (n=15)	3,5%	
DIN 277	Construction time PPP/CPP	-29%	
	Time overruns (n=15)	0%	
DIN 18960 CG 100	Capital costs	-8%	-8%
CG 200	Property management	18%	27%
CG 300	Operating costs	33%	-5%
CG 310	Supply	11%	-10%
CG 320	Waste disposal	0%	0%
CG 330/340	Cleaning	-15%	-14%
CG 350	Maintenance and inspection	0%	0%
CG 360	Energy management	0%	0%
CG 370	Insurance, taxes	211%	-22%
CG 390	Other operating costs*	7310%	5955%
CG 400	Repairs (including risk costs)	69%	-21%
CG 200/350/400	Maintenance budget	80%	-40%
	in % p.a. PPP/KGSt target	1,8%	1,2%
	in % p.a. KGSt actual/BKI	0,6%	1,7%
Usage costs excluding risk costs		11%	-7%
Usage costs including risk costs		10%	-7%
Residual value		30%	0%
Life cycle costs excluding risk costs		39%	-17%
Life cycle costs including risk costs		-17%	-20%
Life cycle costs including risk costs and VAT		-21%	-23%

*Percentage of cost group CG 390 in the usage costs: PPP 3.2% (of which approx. 3% is attributable to CG 200), KGSt: 0.3% BKI: 0.2%

Table 3.1: Results of PPP renovation projects

a) RESULTS:

(190) In comparison to the KGSt key figures, the two PPP renovation projects have higher usage costs excluding and including risk costs (+11%/+10%) as well as higher life cycle costs excluding risk costs (+39%). In comparison, the PPP results for life cycle costs including risk costs (-17%) and including risk costs and additional VAT income (-21%) are positive.

PPP/KGSt
Usage costs
-/+ risk:
+11%/+10%
Life cycle costs +/- risk: +39%/-17%%
and
-21% including VAT

PPP/BKI:Ø -7/-23%

(191) In contrast, the figures for the two PPP renovation projects fall below the BKI key figures for usage costs excluding and including risk costs (-7%) and for life cycle costs excluding and including risk costs (-17%/-20%) as well as for life cycle costs including risk costs and additional VAT income (-23%).

(192) In the sensitivity analysis with the variation of the construction price index using the price increase rate of 2.0%, the PPP advantage is reduced to -9% compared to the KGSt model and to -17% compared to the BKI model. With an annual increase in prices of 1.5%, the PPP advantage falls to -7% (KGSt) and -16% (BKI).

Sensitivity analysis
Indexation of repairs and residual value with 2% and 1.5% p.a.

Indexing residual value and repair costs	PPP/KGSt	PPP/BKI	PPP/KGSt	PPP/BKI	PPP/KGSt	PPP/BKI
Usage costs excluding risk costs	11%	-7%	10%	-8%	9%	-9%
Usage costs including risk costs	10%	-7%	8%	-8%	9%	-9%
Life cycle costs excluding risk costs	39%	-17%	23%	-15%	18%	-10%
Life cycle costs including risk costs	-17%	-20%	-9%	-17%	-7%	-16%
Life cycle costs including risk costs and VAT	-21%	-23%	19%	-28%	-9%	-10%
	Index 2.92% p.a.*		Index 2% p.a.*		Index 1.5 % p.a.*	

*DESTATIS FS 17 Row 4 construction price index, office buildings, annual price increase rate 2007-2021

Table 3.2: Impact of construction price indexing on the overall result

b) NOTES:

(193) In the case of the two pure PPP renovation projects, an in-depth examination of the evaluation of the renovation using an appropriate BKI benchmark was not possible within the scope of this study. This applies especially to the historical conservation project (project 17).

(194) In the historical conservation project, there were cost increases of 3.5% during the construction process. These are not factored into the cost calculation because there is no accessible official information on cost increases over the course of conventional historical conservation construction projects.

4 Results of the new-build and renovation projects
4.1 Summary

European Comparative PPP Study - Germany		GREENFIELD PROJECTS Projects 1-16		BROWNFIELD PROJECTS Projekte 17-18		ALL PROJECTS Projekte 1-18	
		PPP/KGST	PPP/BKI	PPP/KGST	PPP/BKI	PPP/KGST	PPP/BKI
DIN 276 CG 200-700 Construction costs		-17% (-15% / - 20%)		-11%		-16% (-13% / - 19%)	
Cost overruns (n=15)		0,6% (Median 0%)		3,5%		0,7% (Median 0%)	
DIN 277 Construction time		-30%		-29%		-30%	
Time overruns (n=15)		2,3% (Median 0%)		0%		2% (Median 0%)	
DIN 18960 CG 100 Capital costs		-15%	-15%	-8%	-8%	-14%	-14%
CG 200 Object management costs		12%	23%	18%	27%	14%	24%
CG 300 Operating costs		19%	-24%	33%	-5%	20%	-20%
CG 310 Water, heating, electricity		-16%	-33%	11%	-10%	-15%	-30%
CG 320 Disposal costs		0%	0%	0%	0%	0%	0%
CG 330/40 Cleaning costs		5%	9%	-15%	-14%	2%	6%
CG 350 Servicing&inspection		685%	-21%	0%	0%	685%	-21%
CG 360 Energy management		0%	-2%	0%	0%	0%	-2%
CG 370 Taxes, contributions, assurance		179%	-7%	211%	-22%	179%	-11%
CG 390 Other operating costs*		455%	1660%	7310%	5955%	455%	1660%
CG 400 Repair costs		66%	-10%	69%	-21%	68%	-9%
CG 200/350/400 Maintenance budget		137%	-15%	80%	-40%	128%	-17%
in % p.a. PPP / KGST-TARGET		1,6%	1,2%	1,8%	1,2%	1,6%	1,2%
in % p.a. KGST-ACT / BKI		0,6%	1,7%	0,6%	1,7%	0,6%	1,7%
Usage costs without risk costs		-1%	-15%	11%	-7%	0%	-14%
Usage costs including risk costs		-3%	-15%	10%	-7%	-2%	-13%
Residual value		27%	0%	30%	0%	27%	0%
Life cycle costs without risk costs		-3%	-32%	39%	-17%	2%	-30%
Life cycle costs including risk costs		-35%	-34%	-17%	-20%	-32%	-32%
Life cycle costs including risk costs and VAT		-37%	-36%	-21%	-23%	-34%	-34%

* Percentage of the cost group CG 390 in the usage costs: PPP 3.2% (of which approx. 3% costs are attributable to KG 200), KGST: 0.3%, BKI: 0.2%

Table 4.1: Comparison of life cycle costs PPP vs. KGST/BKI -new-build and renovation projects

a) RESULTS:

(195) The usage and life cycle costs for the 18 PPP new-build and renovation projects are almost the same as those of the KGSt model (0%/2%), excluding maintenance risk. The PPP life cycle costs excluding risk costs are slightly lower than the costs of the KGSt model (-2%). The PPP life cycle costs including risk costs and including risk costs and additional VAT income are significantly lower than the KGSt model (-32%/-34%).

PPP/KGSt
Usage costs:
-/+ risk: Ø 0%/-2%
Life cycle costs:
-/+ risk: Ø 2%/-32%
including VAT:
Ø:-34%

(196) The usage and life cycle costs for all 18 PPP projects are between -14% and -34% under the costs of the BKI model.

PPP/BKI -/+ risk
Ø -14%/-34%

(197) In the sensitivity analysis with the variation of the construction price index using the price increase rate of 2.0%, the PPP advantage for life cycle costs including risk assessments is reduced to 24% compared to the KGSt model and to 27% compared to the BKI model. With an annual increase in prices of 1.5%, the PPP advantage falls to 21% (KGSt) and 25% (BKI).

Indexing residual value and repair costs	PPP/KGSt	PPP/BKI	PPP/KGSt	PPP/BKI	PPP/KGSt	PPP/BKI	
Usage costs excluding risk costs	0%	-14%	-1%	-14%	-1%	-14%	Sensitivity analysis
Usage costs including risk costs	-2%	-13%	-2%	-14%	-2%	-14%	Indexation of repairs
Life cycle costs excluding risk costs	2%	-30%	-1%	-25%	-3%	-24%	and residual value
Life cycle costs including risk costs	-32%	-32%	-24%	-26%	-21%	-25%	with 2% and 1.5%
Life cycle costs including risk costs and VAT	-34%	-34%	-26%	-29%	-22%	-27%	p.a.
	Index 2.92% p.a.*		Index 2% p.a.*		Index 1.5 % p.a.*		

Table 4.2: Impact of construction price indexing on the overall result

b) NOTES:

(198) The inclusion of the two PPP renovation projects does not result in any significant change in the results.

4.2 Savings

(199) The percentage PPP advantages are based on corresponding cost advantages.

(200) In the 18 PPP projects analysed, the total PPP construction costs amounted to €616 million, which is €141 million below the BKI key figures[22]. Given that capital costs (repayment of construction costs and interest) account for 60% or more of the usage costs over the term of the contract, construction costs are of particular importance. The efficient construction process lays the foundation for organising maintenance management in particular with sufficient numbers of personnel and financial resources throughout the operation phase.

Construction costs: €141 million savings

European comparative PPP study (Germany)	GREENFIELD PROJECTS + COMBINED GREENFIELD/ BROWNFIELD PROJECTS n=16		BROWNFIELD PROJECTS n=2		ALL PPP PROJECTS n=18	
PPP construction costs	579.107.769 €		36.841.207 €		615.948.976 €	
Difference		-135.222.483 €		-5.415.981 €		-140.638.464 €
PPP usage costs excluding risk costs	1.549.111.276 €		122.571.743 €		1.671.683.019 €	
Difference	-47.212.843 €	-302.274.231 €	13.045.729 €	-9.713.670 €	-34.167.114 €	-311.987.901 €
PPP usage costs including risk costs	1.553.828.718 €		122.571.743 €		1.676.400.462 €	
Difference	-65.425.877 €	-297.556.789 €	11.327.476 €	-9.713.670 €	-54.098.401 €	-307.270.459 €
PPP life cycle costs excluding risk costs	584.933.934 €		44.649.848 €		629.583.782 €	
Difference	-48.928.051 €	-303.989.440 €	13.045.729 €	-9.713.670 €	-35.882.322 €	-313.703.109 €
PPP life cycle costs including risk costs	542.899.483 €		38.638.230 €		581.537.714 €	
Difference	-271.015.746 €	-298.844.229 €	-8.328.426 €	-10.453.234 €	-279.344.172 €	-309.297.464 €
- as above including additional VAT revenu	526.533.349 €		36.547.073 €		563.080.422 €	
Difference	-287.381.880 €	-315.210.364 €	-10.419.583 €	-12.544.391 €	-297.801.464 €	-327.754.755 €

Table 4.3: Construction, usage and life cycle cost savings

(201) The total PPP usage costs for the 18 PPP projects excluding risk costs over approximately 25 years totalling €1.7 billion are €34 million below the KGSt actual values and €312 million below the BKI key figures. If potential damage due to neglected maintenance is considered, the PPP advantage in terms of usage costs including risk costs increases to €54 million in comparison to KGSt.

PPP usage cost savings vs.: KGSt: €34 million to €54 million BKI: €312 million

(202) Comparing public-private partnerships and KGSt, the total PPP life cycle costs excluding risk costs (€630 million) are €36 million lower than those of KGSt, while the PPP life cycle costs including risk costs are €279 million lower than the KGSt model. The difference of €244 million is due to the risk of shorter useful lives as a result of insufficient maintenance budgets, which leads to reduced residual values.

PPP life cycle cost savings vs. KGSt: €-36 million/€- 279 million (exclu- ding/including risk)

(203) Comparing public-private partnerships and BKI, the total PPP life cycle costs excluding risk costs are €314 million lower than those of BKI, while the PPP life cycle costs including risk costs are €309 million lower than the BKI model.

vs. BKI: €-314 million/€- 310 million (exclu- ding/including risk)

(204) The additional VAT income generated by the 18 PPP projects over the 25-year duration of the contracts comes to €18.5 million.

€18.5 million additional VAT income

5. Results of the questionnaire on cost and deadline certainty
5.1 Construction costs

PPP construction cost certainty	Actual/target
Maximum	-1,0%
1st quartile	0,0%
Median (MED)	0,0%
3rd quartile	0,9%
Minimum	5,7%
Average value (AV)	0,9%
Average weighted value (AWV)	1,1%
Ø MED/AV/AWV	0,7%

n=15

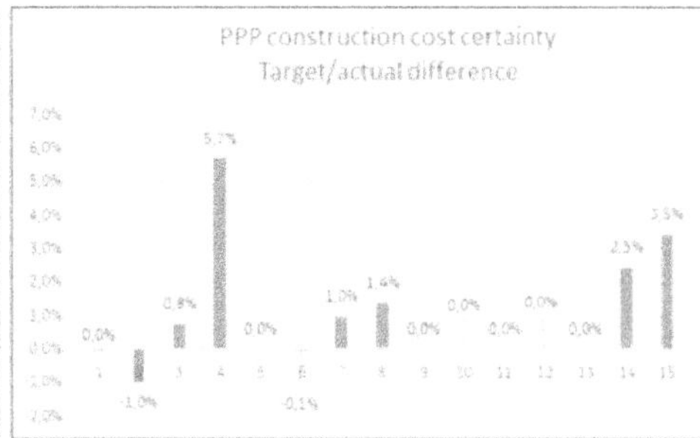

Table 5.1: PPP construction cost certainty

a) RESULTS:

(205) Of the 15 projects with data on this topic, two were settled below the contractually agreed costs. In five projects, there were cost increases of between 0.8% and 3.5%. In one project, the increase came to 5.7%.

High PPP construction cost certainty

(206) The median of the 12 projects is 0%, the average of the mean values is 0.7%. The additional costs are the result of later user requirements (for example, significant enlargement of the originally planned underground car park).

b) NOTES:

(207) The majority of projects show a high level of cost certainty. If additional costs are incurred, this is usually a result of additional user requests: In one project (+5.7%), the number of underground car park spaces was significantly increased during the construction process (from 24 to 170). The project with a 3.5% cost increase is a historical conservation renovation project.

Exceptions: additional requests from users, historical conservation

(208) The contracting parties report that they aim to adhere to the agreed costs from the outset; unforeseen cost increases during the construction phase are offset by optimisations and savings elsewhere.

Cost management for construction

(209) A further reason for compliance with the contractually agreed costs on the part of the local authority is the economic viability analyses carried out beforehand. The economic advantages over conventional benchmarks are a prerequisite for the regulatory authorities and audit offices accepting the project, and possibly also for the granting of subsidies. Therefore, there is a particular incentive in this case to ensure that the contractually agreed costs are honoured.

Incentive structures to comply with construction cost plans

Economic viability studies for project development and contract award

5.2 Construction time
a) RESULTS:

PPP deadline certainty	Target/actual
Minimum	-11%
1st quartile	0%
Median (MED)	0%
3rd quartile	0%
Maximum	45%
Average value (AV)	2%
Average weighted value (AWV)	4%
Ø MED/AV/AWV	2%

n=15

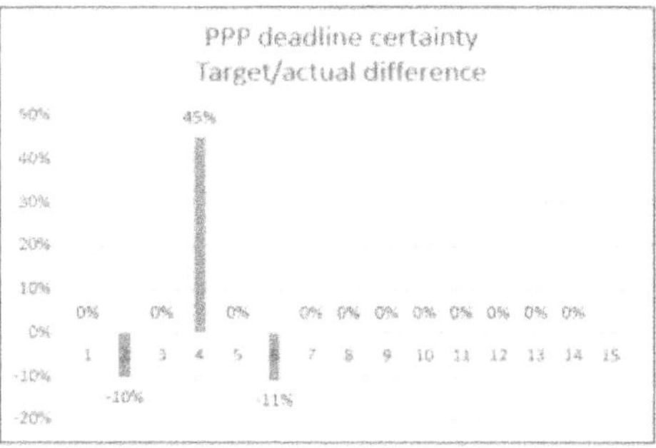

Table 5.2: PPP deadline certainty

(210) 12 of the 15 projects were completed exactly on the contractually agreed date, and two projects were even completed ahead of schedule. In one project, the construction took 45% longer than agreed due to a subsequent

High PPP deadline certainty

user request (significant expansion of an underground car park).

(211) The median of the 14 projects is 0%, the average of the mean values is 2%.

b) NOTES:

(212) As with the construction costs, there are a number of reasons why in PPP projects, construction times, which are often short, are adhered to so effectively.

Incentive structures for meeting deadlines

(213) A key incentive is the agreed fixed price of costs and the resulting interest rate hedging required for interim financing. This means that the PPP company must agree to a schedule for the outflow of funds, which fixes the payment dates for the payment of the loan instalments. Fixed interest rates are agreed with the refinancing bank for the various payment periods of the individual loan instalments when the contract is concluded. The PPP company therefore has a vested interest in ensuring that the schedule is observed so that the project is completed on time and within budget. The interest premium to be paid due to interest rate hedging is therefore directly linked to the agreed cost and schedule.

Interest rate hedging fixes costs and schedule

(214) A further incentive for the PPP company to comply with the agreed deadlines are the agreed contractual penalties for failure to complete the project on time.

Contractual penalties

(215) Furthermore, the interests of the public contractual partner are in alignment, which is a result of the special features of the PPP process with the requirement for economic viability studies to be conducted during the project development and tendering process. The public sector is keen not to contradict the economic viability studies carried out in prior stages, so the risk of additional user requirements arising after the conclusion of the contract is reduced as part of the tender preparation process through consistent user involvement from the outset, which naturally has an impact on adherence to the calculated construction times.

The public authorities' interest in keeping to the schedule

5.3 Cost and deadline certainty in operation

PPP Cost certainty in operation	Target/actual
Minimum	0,0%
1st quartile	0,0%
Median (MED)	0,6%
3rd quartile	1,2%
Maximum	10,0%
Average value (AV)	1,5%
Average weighted value (AWV)	1,3%
Ø MED/AV/AWV	1,2%

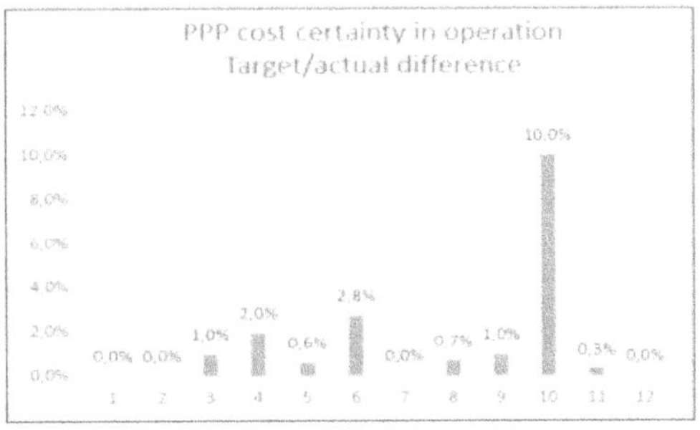

n=12

Table 5.3: PPP cost and deadline certainty in operation

a) RESULTS:

(216) For four of the 15 projects, no changes in performance have been reported to date. In seven projects, there were additional services accounting for between 0.3% and 2.8% of the usage costs. One project resulted in additional benefits of 10% (expansion of a school into an all-day school). The median of the 12 projects is 0.6%, the average of the mean values is 1.2%.

High PPP operating cost certainty

(217) The above figures do not include the savings achieved to date and projected future savings from energy management, which have already been factored into the calculation of usage costs. The achieved and forecast savings account for an average of 0.8% of the usage costs (arithmetic mean).

Savings from energy management

b) NOTES:

(218) In the majority of projects, additional costs beyond the contractual target are the result of additional user requirements.

Reasons for additional costs: Additional services

(219) Problems only arose after use began in one renovation project, where a defect in the building fabric that had not been detected at the time of renovation had to be repaired at the expense of the PPP company.

(220) Reductions in remuneration due to non-compliance with service levels are only reported to occur to a very limited extent.

No reductions in pay at the company to date

5.4 Cost and deadline certainty for conventional projects

(221) Comparable BKI data on cost and deadline certainty for conventional projects have not yet been made available and could therefore not be taken into account in this study[23].

No conventional comparison data

(222) An internet search carried out in December 2020 for "Verzögerungen bei Schulbauen" (delays in school construction) resulted in 48 hits (see Annex B1). The average delay in these projects was seven months. This is 27% of the average construction time for schools, which is 25.9 months.

Internet research on delays to the construction of school buildings:
Ø + 7 months (+27%)

(223) An internet search for "Kostensteigerungen/Kostenexplosion bei Schulbauen" (cost increases/cost explosion for school construction) resulted in 50 hits (see Annex B2) with an average cost increase of 116%. However, it should be noted that the cost increases reported here were incurred not only during the construction phase, but also during the project development phase.

Internet research on exploding costs in school construction:
Ø +116%
But: Project development and construction phase

(224) Frequently cited reasons for conventional cost increases are a long period between the decision to go ahead with the project and the actual construction phase, planning errors, separation of responsibility for planning and implementation, inadequate cost estimates and poor cost management, increases in construction prices, additional user requirements, rescheduling during the construction phase, a lack of experts in construction management and legal proceedings.

Reasons for cost increases

(225) The comparative data for PPP projects from the start of the project development phase could not be examined in the given time available for this study.

Agenda: PPP cost increases from the start of project development?

6 Evaluation of construction and operation by the contracting public authority

European comparative PPP study (Germany) Questionnaire for the evaluation of construction and operational PPP performance																	
PPP-Projekt Nr:	1	2	3	4	5	6	7	8	9	10	12	13	14	15	16	18	Ø
A. Investment phase	1,3	2,0	2,4	1,0	2,4	1,1	1,3	2,0	2,0	1,8	1,7	1,7	1,0	2,0	1,7	2,0	**1,7**
DIN 276 — Construction costs	1	2	3	1	3	1	1	2	2	2	2	2	1	2	2	2	1,8
Construction quality	1	2	1	4	1	1	2	2	2	2	2	2	1	2	2	2	1,8
DIN 277 — Process	2	2	2	1	2	2	2	2	2	3	2	2	1	2		2	1,9
o Call for tender		2	1	1	1	1	2	2	2	2	1	1	1	2	1	2	1,5
o Planning permission		2	1	1	1	1	2	2	2	1	1	1	1	2	1	2	1,4
o Construction phase		2	1	1	1	2	2	2	2	1	1	1	1	2	1	2	1,5
B. Usage phase (DIN 18960)	1,1	2,1	1,4	1,3	1,5	1,2	2,6	2,9	2,9	1,2	2,3	2,3	1,5	2,0	1,2	2,0	**1,9**
KG 100 — Financing	1	2	1	1	1	1	2	2	2		1	1		2	1	2	1,4
KG 200 — Property management	1	2	1	1	1	1	2	3	3	1	2	2	3	2	1	2	1,8
KG 202 — o Caretaking services		2	1	1	1	1	4	3	3	1	2	2	1	2	2	2	1,9
KG 204 — o Einbindung ext. Nachunternehmer			2	2	2		3	2	2	1	2	2	1	2	2	2	1,9
KG 310 — Versorgung, Energiemanagement	1	2	1	1	1	1	2	3	3	1	5	5	1	2	1	2	2,0
KG 320 — Waste disposal	1		2	1	1	1	2			1	1	1	1	2	1	2	1,3
KG 330 — Cleaning	2	3	2	2	2	2	5	4	4	1	2	2	3	2	1	2	2,4
KG 350/400 — Maintenance incl. Repairs	1	2	1	2	1	1	2	3	3	1	3	3	1	2	1	2	1,8
KG 350/400 — o Service levels compliance	1	2	1	1	1	1	2	3	3	1	3	3	3	2	1	2	1,9
KG 350/400 — o Reserve account	1	2	1	1	2	1	2	3	3		3	3		2	1	2	1,9
KG 390 — Cafeteria operation	3		2	1	3	3	3				3	1	1	2		1	2,1
C. Current building condition	1,0	1,0	1,0	1,0	1,0	1,0	2,0	3,0	3,0	1,0	2,0	2,0	1,0	2,0	1,0	2,0	**1,6**
D. Results Evaluation by the PPP municipality	1,1	1,7	1,6	1,1	1,6	1,1	2,0	2,6	2,6	1,3	2,0	2,0	1,2	2,0	1,3	2,0	**1,7**

[3] Quality assessment: 1 = very good / 2 = good / 3 = satisfactory / 4 = sufficient / 5 = unsatisfactory

Table 6.1: Evaluation of construction and operation by public contractors

6.1 Evaluation of individual services in overview

(226) The public contractors were surveyed in 2023 to assess the construction and operating services. In the survey, the individual services from the construction and operating phases were evaluated with scores in addition to four open questions for respondents to answer. Responses were received from 11 municipalities for 16 of the 18 projects within the given timeframe.

(227) Construction costs (1.8) and construction quality (1.8) were rated with an upscale good on average. In one project, the construction quality was poor (4), while the price of this project was very good (1), but there were construction defects.

Construction costs: 1.8 Construction quality: 1.8

(228) The average rating of the PPP method is good (1.9). Tendering (1.5), planning permission (1.4) and the construction phase (1.5) were in fact rated as good to very good.

Process: 1.9 Tendering:1,5, Permission: 1.4, Construction phase: 1.5

(229) Financial performance is rated as good to very good (1.4).

Financing: 1.4

(230) Property management was rated with an upscale good (1.8), as was the involvement of subcontractors (1.9) and caretaking services (1.9). In one case, which was only rated as sufficient, there were coordination problems with the caretakers to be provided by the municipality in accordance with the original contract content. The contractual arrangements have since been changed and the caretakers are now provided by the PPP company.

Property management: 1.8

(231) On average, energy management is rated good (2.0). The average value is negatively affected by the two projects which were rated as "unsatisfactory." The local authority client is clearly dissatisfied with the terms of the contract, according to which the PPP company is only required to honour guaranteed contract volumes and any savings made compared to the guaranteed volumes

Energy management: 2.0

are shared equally. As a result, the PPP company will not implement any additional savings measures at its own expense because the necessary investment costs will not be recouped in the remaining contract period. However, this complaint does not acknowledge that under the PPP ownership model, the local authority is in principle at liberty to have additional energy optimisation measures carried out at its own expense at any time.

(232) Waste disposal services are rated as good to very good (1.3).

(233) In contrast, the average rating for cleaning services is only good to satisfactory (2.4). In one case, there was a poor score (5) because there were disagreements about the need for additional cleaning during the coronavirus pandemic.

Waste disposal: 1.3

Cleaning: 2.4

(234) The cafeteria operations were rated as good (2.1).

Cafeteria: 2.1

(235) Maintenance services are rated as good (1.8), and the same applies to the reserve account (1.9) and compliance with service levels (1.9).

Maintenance: 1.8

(236) The current condition of the building is good to very good (1.6).

Current condition of the building: 1.6

(237) The average rating of the PPP performance for the investment phase, the usage phase and the current condition of the building is an upscale good (1.7).

Overall rating: 1.7

(238) The overall score from the perspective of the municipal contractual partners is therefore once again significantly better than the already good qualitative result (2.2) achieved in this study (cf. Annexes A2 and A4 with the overview of results for the individual projects).

European Comparative PPP Study (GER) Quality of PPP services		EU-Study (GER) Result*	Communal questionnaire
Investment phase		**2,2**	**1,6**
DIN 276	Building standard / building costs + quality	2,7	1,8
	Cost certainty	1,2	
DIN 277	Construct. time+schedule certainty/process	1,0	1,5
	Interest rate hedging	1,0	
Usage phase (DIN 18960)		**2,5**	**1,8**
KG 100	Financing	2,7	1,4
KG 200	Property management	1,8	1,8
KG 310	Supply	2,3	2,0
KG 320	Waste disposal	3*	1,3
KG 330/3	Cleaning	2,9*	2,4
KG 350	Inspection and maintenance	1,4	
KG 370	Taxes and insurances	2,8*	
KG 390	Cafeteria operation	3*	2,1
KG 390	Other operating costs	2,6*	
KG 400	Repairs/maintenance	2,1	1,9
Residual value		**2,1**	**1,6**
Total result		**2,2**	**1,7**

Subjective assessment of the municipalities significantly better than the good qualitative rating in this study

* No in-depth qualitative study to date

Table 6.2: Qualitative assessment of construction and operation - comparison of previous assessment / municipal assessment

(239) This is mainly due to the following aspects:

o The price-performance ratio (construction costs/construction quality) is rated significantly better by the municipal representatives (1.8) than the previous assessment of the construction standard (2.7). Whereas previously only the aspects of construction time and deadline reliability (1.0) were taken into account in the study, the score of 1.5 includes the tendering and planning permission process in addition to the construction time, although the overall assessment of this is in the good to very good range.

o The very good rating for financing performance (1.4) is also striking. In the case of PPP financing, the study gave the projects with project financing a good rating (2), while the projects with forfaiting and waiver of defence were given a medium rating (3) as with the conventional method.

o For the cost groups KG 320 (waste disposal), KG 330/340 (cleaning), KG 370 (taxes and insurance) and KG 390 (cafeteria), there has been no in-depth qualitative analysis to date, which is why a medium rating was generally given here, as with the conventional method.

6.2 Open question: "What appeals to you in particular about public-private partnerships?"

(240) It was repeatedly emphasised that the structural condition of the PPP buildings is superior to that of buildings constructed using conventional approaches. The main reason given for this was the consistent rectification of structural defects and the regular inspection of the buildings and the resulting cosmetic repairs and maintenance work, which enables defects to be rectified quickly. The property is therefore kept in good condition at all times. This also leads to high levels of satisfaction among school management, which is commonly expressed at regular meetings.

> **PPP advantage: good structural condition of PPP buildings because of high-quality maintenance**
>
> **Satisfied users**

(241) Also listed here:

- Trust-based cooperation

- The investment phase, essentially the ownership model

- Completion on schedule in a short construction period

- Independent upkeep, support, maintenance, servicing and inspection by the investor and its subcontractors

- "Everything from one source"

- The current situation on the job market: For example, at the moment it is almost impossible to recruit engineers and architects for property management of schools in the public administration. The PPP contract has enabled a long-term partnership to be established for the management of two large properties for more than a decade. The local authority is now benefiting from the decision made in the past.

> **Trust-based cooperation**
>
> **PPP ownership model, short construction time**
>
> **Personal responsibility of the PPP company**
>
> **Everything from one source**
>
> **Lack of engineers and architects in public administration**

6.3 Open question: "Where is there room for improvement?"

(242) The most frequent area requiring improvement is the operation of the cafeteria (5x): This could be better handled by the Office for School Development (1x). The issue here is that the caterer changed frequently, making the contractors' work more difficult (3x). In one case, better coordination regarding the operation of the cafeteria and implementing quality management user surveys were suggested.

> **Cafeteria operation**

(243) In three projects, cleaning services are identified as in need of improvement: in two projects, the quality of services is impaired by a shortage of staff; in the other case, improvements need to be made to follow-up inspections, particularly for basic cleaning (3x). In one case, there were discrepancies during the coronavirus pandemic due to the additional cleaning services that had become necessary.

> **Cleaning**

(244) With regard to energy management, two projects are criticised because energy investments are strongly influenced by the amortisation period (see above).

> **Contractual provision of energy management**

(245) When it comes to property management, there is a need for better compliance with service levels and better computer-aided facilities management (CAFM) tracking and information transfer.

> **Better management (CAFM)**

(246) In one case, it was stated that the (final) financing should preferably be

> **Final financing via**

provided by the client.

local authority

(247) In one project, more active communication was requested during maintenance. Employees at the local authority are not experts in maintenance and are often unable to assess defects correctly.

Better information management for maintenance

(248) For one project, there was a comment regarding this issue. "PPP projects are more expensive, but pay off over the life cycle. When this cycle is complete, the public sector will be handed a flawless building."

PPP is more expensive, but it is profitable over the life cycle: Faultless building at the contract end

(249) For five projects, no necessary improvements were identified.

6.4 Open question: "What has your experience been like during the pandemic?"

(250) The local authority clients report that there were no apparent problems with operations during the coronavirus pandemic. This is interesting, as the lack of flexibility to react to events that were unforeseeable when the contract was concluded is often cited as a disadvantage of public-private partnerships.

No problems during the COVID-19 pandemic

(251) The individual answers were as follows:

- During the COVID-19 pandemic, there were no problems in operating processes. This is due to the good cooperation between the PPP company and the local authority. The people responsible for the project had already known each other for some time when the pandemic began, so the negative effects could be overcome thanks to this foundation of trust. Constructive solutions were found during frequent telephone calls where preliminary discussions were held on all issues.

Trust-based cooperation

- The operator reacted "flexibly."

- While the school was closed, building services were maintained. There were no issues when work resumed. The ministries' requirements were also met with the assistance of the other contracting party.

Flexible operator response

Ministerial requirements were implemented

- It went very well and people adapted to the new situation quickly.

- Seamless operations were ensured at all times.

Rapid adjustment to new situations

- The necessary measures were implemented reliably. The local authority did not experience any complications and the necessary measures were implemented satisfactorily.

Smooth process

Reliable processing

- In one case, there were disagreements about the additional cleaning requirements.

6.5 Open question: "Would you use public-private partnerships again?"

(252) This question was answered by eleven local authority clients, with seven positive, one negative and one neutral response ("These are political decisions"). Among the positive comments, the challenging staffing situation in the administration was pointed out several times.

7 x yes, 1 x no, 3 x open

(253) In one instance, the following detailed comment was made: "The construction of more schools in the framework of PPP projects is an advantage for the public sector, as carrying out projects of this type involves a large investment in personnel for the public administration and there is often a shortage of staff. The know-how of private companies is invaluable in projects of this kind. The operation of school buildings that have already been constructed by an operator is also a positive development, as there are often no school administrators, facility managers or building technicians in the public sector. It is also clear in the regulations that the installer or operator is responsible for the building."

Main reason: Difficult situation concerning staff in the administration

Clear responsibility for property with PPP companies

(254) Another local authority stated that at least parts of the existing PPP contract could be applied to other schools. In particular, this is due to the current challenges in recruiting specialist staff to manage properties and carry out any

Expenses for processing of PPP contracts

construction work.

(255) In one case, a positive rating was given alongside a note: "even if handling the contract involves much more effort than with the conventional process."

Problem: Incorporation of architectural services

(256) The negative rating was connected with the term "architectural quality."

7 Overview of the cost and quality results
7.1 New-build projects
7.1.1 Construction
7.1.1.1 Construction costs

(257) The construction costs for PPP projects are on average 17% lower than the BKI key figures (between 15% and 20%). The buildings are constructed to an average and in some cases above-average standard.

Construction costs: Ø -17% < BKI

(258) The projects demonstrate a high level of cost certainty: The contractually agreed costs are largely honoured, and in some cases are even undershot. The average cost increase is 0.5%, with a median of 0% (n=14).

High cost certainty

7.1.1.2 Construction time

(259) The PPP construction time is on average 30% lower than the comparable BKI values. Some of the large projects with up to four locations were completed in the same time it would take to build a conventional project.

Construction time: Ø -30% < BKI

(260) The level of deadline certainty in public-private partnerships is high: Most projects were completed precisely within the agreed schedule, and two projects were even completed ahead of schedule. In one project, there was a significant delay as a result of a significant subsequent increase in the number of underground car parking spaces. Deadlines are exceeded on average by 2.3%, with a median of 0% (n=12).

High deadline certainty

7.1.1.3 Qualitative aspects: Efficient construction process and life cycle approach

(261) Construction cost savings of 15-20% over comparable conventional BKI values with average to above-average quality and remarkable cost certainty, as well as 30% shorter construction times with a high degree of deadline certainty — these are the hallmarks of the PPP method's highly efficient planning and construction process.

(262) There are striking organisational differences between this and the conventional process, with bids placed by individual trades. With public-private partnerships, the PPP company is responsible for the planning and construction process. It bears corporate responsibility not only during the construction phase, but over the entire life cycle, with all the associated potential risks and rewards. The PPP company therefore has a vested interest in delivering good construction results, as this is the foundation for a successful long-term contractual relationship with adequate profit expectations and a satisfied contract partner. Conversely, in the conventional process this type of interlinking of personal responsibility with potential risks and rewards for those involved in planning, construction and operation over a comparatively long life cycle does not exist, or at least does not apply to the same extent.

Differences in organisation compared to the conventional procedure

PPP companies are responsible for their own potential risks and rewards over the life cycle

Efficient construction processes are fundamental for successful operations

(263) Given the corporate responsibility for construction and operation, PPP planning is focused on the life cycle approach. The positive effects are now clearly evident in maintenance and energy management (see below).

Life cycle planning for construction and operation

(264) There are further reasons for the high level of cost and deadline certainty in the construction process: The subcontractors engaged by the PPP company also have a strong vested interest in ensuring that their construction services are provided on time and in accordance with the contract. The incentive for these companies to perform well is that they remain part of the PPP company's construction team and are therefore able to guarantee that maintenance services are carried out on site during the operating phase with short response and rectification times. This also increases the chances of

PPP construction team: Personal interest in cost and deadline certainty

being contracted by the PPP company for other projects. All of this helps to secure long-term order volumes and has a remarkably positive impact on small and medium-sized enterprises, as studies conducted as part of the PPP pilot project Südbad Trier have shown[24].

Service levels increase SME friendliness

(265) The fixed price of the construction costs means that the interim financing costs are also fixed; if the deadlines are missed, the entire financing plan is at risk. This is another reason why PPP companies are willing to resolve any issues that arise during the construction process quickly and without halting construction, obtaining expert opinions or initiating legal proceedings (see the example of Südbad Trier[25]).

Interim financing and fixed construction cost price

(266) The PPP company also takes the construction price increase into account when calculating the fixed construction cost price. Taking construction price increases during the construction phase into account is not permitted in conventional cost planning for budgetary reasons.

Fixed construction costs and price increases

(267) PPP projects also often stipulate contractual penalties for late completion. The performance in terms of construction time was very good in all the projects analysed here. Whether and to what extent this aspect is relevant could not be investigated in depth in this study.

Contractual penalties for late delivery

(268) The high level of cost and deadline certainty in the construction process can also be seen in connection with the special features of the PPP process. During project development, the focus is on carrying out economic viability studies across the planned life cycle period, with the assistance of external technical, business and legal experts. The requirements, quality and upper cost limits are defined here. Involving users early in the project development process also helps to minimise additional user requirements arising during the construction process. The parties involved on the client side therefore have a keen interest in ensuring that the agreed cost limits, quality levels and deadlines are adhered to, in no small part to protect themselves from objections from the audit offices. The PPP company is also aware of this. If problems occur in the construction process, which cannot be completely avoided even with the PPP approach, all parties involved have a common interest in resolving these problems constructively, even if, for example, it is a question of keeping within cost ceilings by making savings elsewhere.

Project development according to FMK Guidelines

With focus on the life cycle approach

Drives the parties' interest in cost and deadline certainty

7.1.2 Financing
7.1.2.1 Costs

(269) PPP capital costs are on average 15% lower than conventional capital costs due to the construction cost advantage. This reduces the construction cost advantage slightly because of higher PPP interest rates (Ø +0.63% for interim financing, Ø +0.18% for projects with forfeiting and waiver of defence and Ø +0.59% for projects with project financing).

Capital costs Public-private partnerships/convent ional procurement processes Ø -15%

Higher interest rates for public-private partnerships

(270) It should be noted that project financing generally incurs additional costs as part of other operating costs (DIN 18960 CG 390).

7.1.2.2 Qualitative aspects

(271) The interest premiums determined from the PPP price sheets are partly explained by differences in risk hedging. The PPP construction costs are fixed prices, so the PPP company must hedge the interest rate risk during the construction period.

Higher interest — more in return? Fixed construction cost price necessitates interest rate hedging

(272) For project financing, the interest conditions include risk premiums for project risks because the loan granted to the PPP company is also at risk if the PPP company performs poorly. The refinancing bank therefore has a vested interest in ensuring the PPP contract is properly fulfilled and therefore conducts independent controls before the contract is concluded and during the course of the project.

Project financing: Quality assurance by the bank

(273) These qualitative differences to the conventional method were not evaluated quantitatively.

7.1.3 Property management
7.1.3.1 Costs

(274) PPP property management costs are 12% higher than KGSt and 23% higher than BKI figures. The actual PPP property management costs will be approximately 3% higher because the relevant cost elements are often recorded in the other operating costs cost group (DIN 18960 CG 390) (see below paragraph 121 f).

Property management:
PPP/KGSt: Ø +12%
PPP/BKI: Ø +23%

(275) These additional PPP costs are neutralised by the fact that the personnel costs of the PPP company are subject to VAT, whereas no VAT is charged on the costs of employing public administration staff. PPP projects therefore generate additional VAT income for federal, state and local authorities.

VAT on PPP personnel — difference from conventional procurement process

7.1.3.2 Qualitative aspects

(276) PPP property management costs of 6.7% (plus a share of other operating costs) are the third largest cost item in the usage costs after capital costs and repairs. For KGSt, property management costs are only in fifth position (6%), for BKI they are in sixth position (4.6%).

Different PPP priorities
Personnel costs
Third largest cost item

(277) This means that PPP companies prioritise the distribution of cost budgets differently to KGSt and BKI. This can be explained by the incentive and liability structures intrinsic to public-private partnerships: PPP companies have a long-term opportunity to make a profit, but in return they have a performance obligation, have to bear cost risks and operator liability. Any deviations from the target service levels may result in fee reductions and claims for compensation. Good performance leads to calculated profit and participation in savings (for example, in energy management and maintenance). It is therefore necessary and beneficial to address this with sufficient and well-qualified personnel and a well-organised company hierarchy with project managers on site and coordinating divisional managers below the management.

Property management is a much higher priority in the budget for public-private partnerships

Cause: Incentive and liability structures

Securing profits and minimising risks require sufficient qualified personnel

(278) In the PPP company's own interest, this incentive and liability structure also requires the establishment and operation of an efficient IT-based controlling and monitoring system to manage operating processes and to minimise risks arising from operator liability. Naturally, the public contracting party also benefits from this.

Efficient controlling and monitoring system

(279) It is logical that the remuneration system also reflects this incentive and liability structure and provides for a performance-related remuneration component in addition to the basic remuneration, which is also linked to an individual's personal contribution to the success or failure of the project. This is extremely unusual in the public sector. In this respect, it is not surprising that the goal of delivering construction work on schedule and within budget will not be achieved if this goal is not supported by incentive structures embedded in the system itself.

Performance-based remuneration system

7.1.4 Energy supply
7.1.4.1 Costs

(280) The expected PPP utilities costs are on average 16% below the KGSt figures and 33% below the BKI figures.

Energy supply:
PPP/KGSt: Ø -16%
PPP/BKI: Ø -33%

(281) PPP contracts typically stipulate that the PPP company guarantees maximum consumption volumes for heat, electricity and water. Should these guaranteed volumes be exceeded, the PPP company bears the cost risk and savings are usually shared equally.

(282) Information on the actual consumption volumes was available for ten of the projects analysed; the estimated PPP supply costs could therefore be determined using the actual consumption volumes and the values for the remaining years of operation derived from this. As a result, the maximum consumption volumes and supply costs will be undershot by an average of 11%, which corresponds to expected savings of €7.9 million across ten

Ø 11% savings compared to guaranteed maximum consumption quantities for heating, electricity and water

projects.

7.1.4.2 Qualitative aspects

(283) When comparing the heat consumption values, the maximum consumption volumes guaranteed with public-private partnerships are 8% below the VDI 3807 guideline, 38% below the average value of VDI 3807 and 36% below the comparable KGSt values.

PPP heating rates
Ø PPP max: -8% <
VDI guide value
-38% < VDI average
-36 < KGSt

(284) The actual consumption values of the PPP projects are 38% below the VDI guideline value, 58% below the VDI mean value and 56% below the comparable KGSt values.

Ø PPP actual:
-38% < VDI guide
value
-58% < VDI average
-56 < KGSt

(285) As a result, it can be seen that the guaranteed maximum PPP consumption volumes were set quite ambitiously and were then significantly undercut by the actual consumption values

Good qualitative
result

(286) In this context, the findings from the analysis of EnEV energy performance certificates to reduce transmission heat losses, which emerged from a master's thesis on PPP energy efficiency, are particularly interesting. The analysis revealed a strong correlation between the contractual provisions on risk sharing for both surplus consumption and under-consumption: The higher the agreed savings participation of the PPP company, the greater the reduction in transmission heat loss: In the case group without savings sharing, the EnEV targets were undershot by 36%; in the case group where half of the savings were shared, this figure was 45%. Where the PPP company was able to retain 100% savings, the EnEV specifications were clearly undershot by 69%. The incentive of receiving 100% of the benefits from future energy savings obviously motivated the PPP company to pay particular attention to reducing transmission heat losses (for example, by improving the external insulation).

Significantly below
EnEV transmission
heat loss
requirements

Correlation between
savings
participation/reductio
n of EnEV target

at 0%: -36%
at 50%: -45%
at 100%: -69%

7.1.5 Waste disposal costs

(287) Due to the data for PPP and BKI being unclear, the KGSt key figures for the disposal of wastewater, rainwater and waste were also used for PPP and BKI. Disposal costs account for 1% (PPP and KGSt) and 0.8% (BKI) of the usage costs.

Waste disposal:
KGSt cost approach
for PPP and BKI

7.1.6   Cleaning
7.1.6.1 Costs

(288) The projected PPP cleaning costs are on average 5% higher than the KGSt figures and 9% higher than the BKI figures.

Cleaning:
PPP/KGSt: Ø +5%
PPP/BKI: Ø +9%

7.1.4.2 Qualitative aspects

(289) A more detailed analysis of cleaning performance was carried out for one of the 16 projects. In comparison to DIN 77400, the performance profile here is 30% more varied and includes additional cleaning services; the annual cleaning area is significantly larger than the KGSt key value. The PPP disadvantage when comparing cleaning costs per m2 of gross floor area therefore becomes a PPP advantage when comparing cleaning costs per m2 of annual cleaning area.

Example cleaning
analysis
Comparison with DIN
77400 and KGSt

(290) Moreover, as part of the survey on cost and deadline certainty, fee reductions as a result of complaints about cleaning services were reported only rarely and only to a minor extent.

Minor deviations
between target/actual

7.1.7 Taxes and insurance

(291) The percentage variations in PPP costs for taxes and insurance compared to the KGSt key figures are +179% and -7% compared to BKI. These costs account for 0.9% (PPP), 0.4% (KGSt) and 1.1% (BKI) of the usage costs. These costs are only partly shown separately in the PPP contracts and then included in other cost groups.

Taxes, among
others:
PPP/KGSt: Ø +179%
PPP/BKI: Ø -7%

7.1.8 Other operating costs

(292) The other operating costs cost group for public-private partnerships shows significant percentage differences compared to KGSt (+455%) and BKI (+1660%). These costs account for 3.2% (PPP), 0.3% (KGSt) and 0.2% (BKI) of the usage costs.

Other costs: PPP/KGSt: Ø +455% PPP/BKI: Ø +1660% But: low share of total costs

(293) Again, these costs are only partly shown separately in the PPP contracts. The high percentage variance compared to KGSt and BKI is due to cost components from other cost groups (such as overhead costs, which are allocated to property management costs) also being recorded here.

PPP: Pro rata allocation to other cost groups

(294) In the case of projects with project financing, this item includes additional ongoing costs incurred by the project company for auditing and tax advice, costs for guarantees and letters of intent as well as external experts engaged for ongoing due diligence (for example, monitoring compliance with service levels).

Special costs for project financing

7.1.9 Maintenance
7.1.9.1 Costs

(295) PPP maintenance costs (i.e. the pro rata personnel costs incurred for maintenance, maintenance and inspection costs and repair costs) are on average 137% higher than the KGSt figures and 14% lower than the BKI figures.

Maintenance: PPP/KGSt: Ø +137% PPP/BKI: Ø -14%

(296) The differences in maintenance and inspection costs are particularly significant: Here, PPP costs are 685% higher than the KGSt figures and 21% lower than the BKI figures. One of the key reasons for these significant differences compared to the KGSt values is likely the different incentive and liability structures of the PPP contract (see below).

Maintenance and inspection PPP/KGSt: Ø +685% PPP/BKI: Ø -21% Major difference from KGSt: PPP incentive and liability system

(297) The PPP repair costs exceed the KGSt figures by 66% and remain 10% below the BKI figures.

Repairs: PPP/KGSt: +66%; PPP/BKI: -10%

(298) The average annual PPP maintenance budget is 1.6% of the restoration costs[26], which is significantly more than the conventional actual budget of 0.6% and slightly less than the BKI budget (1.7%).

Maintenance budget as a percentage of restoration costs

7.1.9.2 Qualitative aspects

(299) According to the results of the PPP school study (2019), only a very low level of maintenance can be achieved with a maintenance budget of 0.6% of the restoration costs per annum; the target range for a medium level of maintenance over 25 years is 1.2% per annum, while a budget of 1.6% or 1.7% enables a high level of maintenance.

PPP school study: 0.6% low level 1.2% average level 1.7% high level

(300) To date, PPP maintenance therefore differs significantly from conventional practice: Maintenance budgets in the target range for a medium to high level of maintenance, improved staffing levels, user requirements for compliance with response and correction times, reserve accounts for rapid access to funds, which the building administration described as a "quantum leap." The principal causes of these differences are different organisational priorities and contractual incentive structures.

Significant advantages in PPP maintenance management

(301) The upper cost limits agreed in the contract for maintenance costs, the definition of qualitative target conditions for key components, and the risk of fee reductions for failure to meet the specified service levels, through to sharing any remaining credit balance in the reserve account at the end of the contract mean that PPP companies must implement strict risk management procedures. This includes the realisation of part-specific maintenance calculations in particular. No offer can be made without this, and the budget that is determined from this is a prerequisite for the conclusion of the contract, resulting in expert knowledge and cost transparency.

Cause: Contractual service and incentive system

Risk necessitates detailed cost calculation

Qualitative, project-focused budget calculation

(302) In contrast, maintenance budgets for conventional projects are generally

Conventional procurement

driven by the budget and are not specific to the property. A survey of the 300 largest German cities in 2020 revealed, among other things, that the competent administrative bodies do not yet carry out part-specific maintenance calculations. 79% of respondents were unable to state how high their maintenance budget is as a percentage of restoration costs. This indicates a lack of cost transparency.

process: no project-focused budget calculation

No detailed cost calculations, no cost transparency

(303) A further major reason for the organisational differences may also be operator liability. In order to eliminate the associated risks not only for the company, but also for the personal criminal liability of the employees, the inspection and maintenance schedules are carried out and documented meticulously. As shown by the considerable differences in maintenance and inspection budgets, this is clearly handled completely differently under the conventional model.

Operator liability necessitates prioritisation of maintenance and inspection

7.1.10 Comparison of usage costs including and excluding risk costs

(304) Despite higher costs for maintenance, property management, cleaning and other operating costs and higher financing interest rates, the PPP usage costs excluding risk costs are 1% below the KGSt model costs. The BKI key figures were undershot by 15%. This is made possible by efficient construction processes and savings in energy management.

Usage costs excluding risk costs: PPP/KGSt: Ø -1% PPP/BKI: Ø -15%

(305) For usage costs including risk costs, 2% of the production costs for the KGSt model were allocated for the risk of structural damage due to insufficient maintenance budgets. The advantage of the PPP usage costs including risk costs therefore increases to 3% compared to the KGSt model. It should be noted that this figure is relatively low (see paragraph 153 above, practical example of the city of Neuwied where structural damage was identified in 16 schools resulting from neglected maintenance which amounted to 7% of the restoration costs).

Usage costs including risk costs: PPP/KGSt: Ø -3% PPP/BKI: Ø -15%

Risk of building damage because of neglected maintenance

7.1.11 Residual value

(306) Based on the results of the PPP school study (2019), it is clear that the effects of the maintenance strategy on the residual value must be taken into account in economic viability studies. As such, the liquidity advantage of low maintenance budgets is offset not only by reduced usage quality during the operating period, but also by structural damage, shortened useful lives and consequently reduced residual values at the end of the comparison period. At the same time, above-average maintenance budgets will lead to an extension of the useful life. It seems to be certain that a connection exists but a quantitative evaluation requires an estimate based on sufficient data. The fact that the useful life is reduced by 30% when maintenance budgets are 50% below target is empirically substantiated by the results of the PPP school study (see above paragraph 162 f).

Inclusion of the prospective residual value in economic viability studies

Correlation between maintenance budget and useful life/residual value

(307) In addition to the interdependence between the maintenance budget and useful life, hidden reserves (see paragraph 164 above) and indexing must also be factored into the residual value calculation (see paragraph 165).

Allowance for hidden reserves and indexing

(308) On this basis, the projected residual value of the PPP projects is on average 27% higher than the KGSt alternative and on a par with the BKI alternative[27].

Residual value: PPP/KGSt: Ø +27% PPP/BKI: Ø +/-0%

7.1.12 Comparison of life cycle costs including and excluding risk costs

(309) The PPP life cycle costs excluding risk costs (i.e. usage costs minus residual value) are on average 3% lower than the KGSt model and 32% lower than the BKI model[28]. The costs of eight PPP projects remain below the costs of the KGSt model, and all PPP projects are below the BKI model.

Life cycle costs excluding risk costs: PPP/KGSt: Ø -3% PPP/BKI: Ø -33%

(310) Life cycle costs, including risk costs, include the maintenance risk and the impact of maintenance budgets on the useful life and residual value. This increases the PPP advantage in terms of life cycle costs including risk costs to 35% over the KGSt model. Compared to the BKI model, the PPP advantage is 34%. The costs of all 16 PPP projects remain below the costs of the KGSt

Life cycle costs including risk costs: PPP/KGSt: Ø -35% PPP/BKI: Ø -34%

model and below the costs of the BKI model.

7.1.13 Factoring in the additional VAT income

(311) The additional VAT income from the PPP personnel costs improves the PPP result over the KGSt and BKI models by approximately 2% of the life cycle costs including risk costs.

PPP leads to additional VAT income

7.1.14 Impact of indexing on the result

(312) The level of indexing has a significant impact on the overall result, particularly with regard to the construction price index, which is used to index maintenance costs and the residual value. From 2007 to 2021, construction prices rose by 2.92% per annum; if the increase in construction prices is adjusted to 2% or 1.5%, the following picture emerges.

Sensitivity analysis Construction price index with 2%+1.5% instead of 2.92% Inflation increases residual value and PPP advantage

(313) If indexed at 2% per annum, the PPP advantage in terms of life cycle costs including risk costs is reduced from 35% to 26% compared to the KGSt model and from 34% to 28% compared to the BKI model.

Index 2% p.a. PPP/KGSt: Ø -26% PPP/BKI: Ø -28%

(314) With an annual increase in construction prices of 1.5%, the PPP advantage is reduced to 23% (KGSt) and 27% (BKI).

Index 1.5% p.a. PPP/KGSt: Ø -24% PPP/BKI: Ø -28%

7.2 Renovation projects

(315) The construction costs for the two renovation projects are 11% lower than BKI. The construction time saving is 29%. It should be noted that it was only possible to compare the PPP renovation performance with BKI key figures to a limited extent.

Construction costs: Ø -11% Construction time: Ø -29% < BKI

(316) The property management costs are 18% higher than KGSt and 27% higher than BKI. The utilities cost figures are worse than for new builds (+11% compared to KGSt, -10% compared to BKI). The cleaning costs are lower at -15% (KGSt) and -14% (BKI). The PPP maintenance budget, expressed as a percentage of restoration costs, is 1.75%, which is slightly above BKI (1.72%) and again significantly above the KGSt key figures (0.63%).

(317) As a result, the PPP usage costs excluding and including risk costs are higher than the KGSt model (+11%/+10%) and lower than the BKI model (-7%).

Usage costs PPP > KGSt (+11/10%) PPP < BKI (-7%)

(318) The PPP life cycle costs excluding risk costs are 39% higher than the KGSt model, including risk costs they are 17% lower. Compared to the BKI model, the PPP life cycle costs excluding and including risk costs are 17% and 20% better, respectively. including additional VAT income, the PPP advantage increases to 21% (KGSt) and 23% (BKI).

Life cycle costs PPP/KGSt: +39/-17% PPP/BKI: -17/-20%

(319) In the sensitivity analysis, the PPP advantage in terms of life cycle costs including risk costs is reduced to 9% (KGSt) or 17% (BKI) with a construction price index of 2% per annum and to 7% (KGSt) or 16% (BKI) with a construction price index of 1.5% per annum.

Sensitivity analysis indexing: PPP advantage declines as inflation falls

7.3 New construction and renovation projects

(320) Usage costs excluding and including risk costs as well as life cycle costs excluding risk costs for the 18 PPP new-build and renovation projects are almost the same as for the KGSt model (0%/-2%/2%). The PPP life cycle costs excluding risk costs are slightly higher than the KGSt model (2%), but including risk costs they are significantly lower than the KGSt model (-32%). Including additional VAT income, the PPP advantage increases to 34%.

PPP/KGSt: Usage costs -/+ risk: 0%/-2% Life cycle costs -/+ risk: 2%/-34%

(321) The usage and life cycle costs for all 18 PPP projects are between 14% and 32% under the costs of the BKI model. Including additional VAT income, the PPP advantage increases to 34%.

PPP/BKI: Usage costs -/+ risk: -14%/-13% Life cycle costs -/+ risk: -30%/-34%

(322) In the sensitivity analysis for indexing maintenance costs and residual value, the PPP advantage in terms of life cycle costs including risk costs is reduced to 24% (KGSt) or 26% (BKI) with a construction price index of 2% per annum,

Sensitivity analysis indexing: rising inflation increases PPP advantage

and to 21% (KGSt) or 25% (BKI) with a construction price index of 1.5% per annum.

8 Conclusion and recommendations

(323) This study shows that the recent debate on the economic viability or economic ineffectiveness of PPP projects has largely been conducted without a reliable data basis. The results of this work are as follows:

o A PPP advantage of 17% to 35% (on average 32%) with savings of €298 to €328 million in life cycle costs over 25 years, taking into account future residual values;

o a highly efficient construction process with 15-20% lower construction costs with medium to good quality, 30% shorter construction times and cost and deadline certainty close to optimal;

o clear qualitative advantages, particularly in property organisation, maintenance and energy management;

o foreseeably significantly better residual values

o and a remarkably good quality assessment of PPP construction and operating services to date (1,7).

They prove that the PPP life cycle approach has so far actually been able to generate the anticipated efficiency potential in the 18 PPP projects examined at the beginning of the German PPP initiative and thus legitimise the implementation of new PPP projects[29].

Good results legitimise new PPP projects

(324) The purpose of economic viability studies for public infrastructure projects is always to examine existing structures for potential improvements and to provide stimulus for the modernisation of public administration.

Stimulus for modernisation of administration

(325) As with the PPP school study (2019), this study shows that there is considerable potential for optimisation in conventional maintenance approaches. Achieving comparable structures would require the following:

Improve maintenance through:

- Introduction of part-specific maintenance calculations during design planning.

Part-specific maintenance calculation

- Definition of service levels for the key building parts with response and rectification times.

Service levels

- Establishing (virtual) reserve accounts allowing funds to be accessed at short notice without lengthy authorisation procedures.

Reserve account

- Introduction of incentive mechanisms such as

  o Sanctions for not meeting user requirements for compliance with response and correction times. This would involve the introduction of fee structures within the administration, according to which the school administration would be entitled to have defects rectified promptly and non-compliance would result in a reduction in fees at the expense of the building and property administration.

Sanctions for not meeting user service level requirements

  o Results-focused remuneration in construction and property management.

Result-focused remuneration

- Legal establishment of minimum maintenance budgets and linking of investment grants to compliance with minimum maintenance budgets; as the private sector incentive mechanisms of PPP contracts (opportunity for profit and risk of loss) are not easily replicated in public structures, the alternative is to fall back on legal requirements or administrative regulations.

Legal minimum maintenance budget

- Legal obligation to report the maintenance indicator "annual maintenance budget as a percentage of restoration costs" in public budgets: As part of the PPP school study, the indicator developed by the KGSt proved to be a suitable indicator for budget planning at a medium maintenance level.

Obligation to report maintenance costs as a percentage of annual restoration costs in budget planning

Efforts should be made to ensure that this indicator is presented transparently in the budget planning for each construction project in order to ensure that the supervisory bodies/authorities can monitor it effectively.

- Determining the future residual value based on the maintenance budget and taking into account hidden reserves and indexation, corresponding supplement in the FMK Guidelines (2006). *(Allowance for future residual values)*

- Changing final financing from full amortisation to partial amortisation for conventional procurement processes and public-private partnerships as it was the case in the 1970s with the first generation of property leasing contracts (full amortisation contracts) and the second generation (partial amortisation contracts[30]. PPP projects with final financing often provide for full amortisation over the duration of the contract in accordance with the client's specifications in the tender documents. In conventional local authority financing practices, long-term financing is usually also fully amortised over 20 to 30 years. As already discussed in the PPP school study, this practice is worth examining if maintenance budgets are too low due to cash flow bottlenecks in the local authority's budget. This is because, according to the golden balance sheet rule, infrastructure should be financed synchronously with the depreciation in value due to use. As shown in the PPP school study (2019), partial amortisation up to 80% (instead of 100%) can generate the necessary liquidity to increase maintenance budgets from 0.6% to 1.2% of the restoration costs per annum[31]. This is naturally offset by a nominally higher level of debt at the end of 25 years, although this is neutralised by the loss of purchasing power due to inflation. *(Partial amortisation for financing creates liquidity for better maintenance)*

- The cash value method is problematic for the elimination of the shortcomings of low maintenance budgets and the resulting reduced quality, shortened useful lives and reduced residual values. Low maintenance budgets (50% below target) create a relative advantage over target budgets in the cash value analysis. The disadvantage of lower residual values is also reduced by discounting. This effect was of minor importance in recent years when capital market interest rates were low, but this will change as interest rates are rising again. The topic therefore appears to be of considerable economic relevance, as insufficient maintenance budgets are likely to account for a significant proportion of the local authorities' cumulative maintenance backlog of €186.1 billion (KfW-Kommunalpanel 2024[32]). *(Limitations of the cash value method for economic viability studies)*

- The breakdown of costs in the PPP tender documents and contracts is not standardised, which makes them difficult to compare. The aim should be to create uniform structures. *(Standardised cost breakdown)*

(326) The results of this study on the importance of incentive structures in reducing energy consumption suggest that such structures should be established more firmly in future. This discussion must also be held with the audit offices, as they have criticised the fact that PPP companies participate in the savings achieved, arguing that this is a waste of public funds because the administration could achieve 100% of the savings on its own if it were organised optimally. *(Incentive structures for energy saving)*

(327) Given that savings of between €298 and €328 million have been achieved in 18 PPP projects over 25 years, the question arises as to whether these results are representative of the 307 German PPP projects which have been carried out to date in the building construction and transport sub-sectors, with construction costs of €17.4 billion and estimated usage costs of €46 billion over 25 years[33]; based on this, calculated savings of over €8 billion could be achieved. If one also considers the potential for optimising the modernisation of conventional (and PPP) structures, it is clear how important evaluations of the economic efficiency of public-private partnerships, research and development and the transfer of knowledge of the PPP life cycle approach are — not least in order to counter the main criticism that the economic viability of public- *(PPP savings potential / For 18 projects approx. €330 million/for 300 projects €8 billion? / Improvement in the transparency of PPP efficiency required / through evaluation, research and development and knowledge transfer)*

private partnerships lacks transparency, a complaint which has arisen repeatedly from audit offices and in the media over the last ten years.

(328) Therefore, to further improve the transparency of the PPP life cycle approach, the following appears advisable:

- Further evaluation work, inclusion of additional sub-sectors

   Evaluation

- More efficient provision of key data on construction and operation; harmonisation of cost breakdowns in tenders and contracts with DIN 18960

   More efficient data transfer

- Improved publication of project and research results (creation of project documentation, update and revision of the PPP project database)

   Better publication of results

- Conducting research on the PPP life cycle approach (for example, comparing financing conditions considering qualitative differences; development of the reserve account; management of vandalism damage; evaluation of annual cleaning areas; impact of indexing provisions with interval periods (such as adjustment for price changes only every five years)

   Further research topics

- Better organisation of knowledge sharing, research and development in a cooperative network of chairs at various universities and colleges focusing on individual key topics (such as the education sub-sector and roads); creation of a cross-sector PPP website, establishment of a PPP academy

   Better organisation

- Financing of knowledge sharing, research and development primarily through the allocation of public funds. This would only require a fraction of the savings made.

   Financing from PPP savings

(329) Article 91d of the German Basic Law establishes a constitutional framework for benchmarking work, and the relevant federal and state budgetary regulations also unanimously require regular performance reviews to be carried out[34]. The fact that the PPP data were predominantly provided by the private sector is noteworthy in this regard. However, Article 91d and the public budget regulations are aimed at policymakers and the administration. It is hoped that this study may help to generate the necessary political will[35].

   The framework for improved benchmarking exists

   Is there any political will for it?

(330) In addition, when discussing public-private partnerships in the future, we should perhaps bear in mind that the two professors of economics Oliver Hart and Bengt Holmström were awarded the Nobel Prize in Economics in 2016 for their work on incentive structures according to the principal-agent theory in incomplete contracts. This also includes PPP contracts. This study confirms that suitable incentive and liability measures in PPP contracts can indeed unlock considerable efficiency potential and that corresponding incentive structures should also be established in conventional procedures as far as possible.

   PPP life cycle approach and incentive mechanisms

   Research focus of the 2016 Nobel Prize in Economics (Hart/Holmström)

ANNEX

A1 Overview of results — Cost comparison of 18 PPP projects/KGSt/BKI

European comparative study on the economics of public-private partnership life cycle costs (Germany)

PPP project no: / comparison with	1 KGSt	1 BKI	2 KGSt	2 BKI	3 KGSt	3 BKI	4 KGSt	4 BKI	5 KGSt	5 BKI	6 KGSt	6 BKI	7 KGSt	7 BKI	8 KGSt	8 BKI	9 KGSt	9 BKI
DIN 276 Construction costs		-21%		-18%		-32%		-6%		-27%		-21%		-9%		-24%		-21%
DIN 277 Construction time		-36%		-35%		-19%		-38%		-35%		-23%		-35%		-41%		-35%
DIN 18960 CG 100		-19%		-17%		-26%		-3%		-23%		-20%		-9%		-22%		-20%
CG 100																		
CG 200	36%	32%	17%	24%	13%	30%	52%	76%	6%	29%	10%	4%	110%	145%	-36%	-33%	-17%	-12%
CG 300	18%	-32%	55%	4%	17%	-16%	30%	-35%	62%	8%	72%	19%	2%	-26%	20%	-33%	18%	-32%
CG 310	-61%	-68%	-62%	-63%	13%	-17%	-33%	-53%	91%	59%	-13%	-25%	-13%	-37%	-38%	-38%	-41%	-41%
CG 320	0%	0%	0%	0%	0%	0%	0%	0%	0%	0%	0%	0%	0%	0%	0%	0%	0%	0%
CG 330/40	49%	-3%	33%	26%	-27%	-20%	58%	74%	3%	6%	21%	76%	-32%	-26%	19%	11%	25%	18%
CG 350	658%	16%	1663%	88%	627%	-11%			757%	-27%	591%	-35%	716%	1%	655%	-59%	478%	-68%
CG 370	1%	-68%			159%	-36%			331%	11%	558%	282%						
CG 390					-76%	-20%	1394%	5173%			317%	15767%			113%	-70%	1210%	80%
CG 400	-20%	-46%	25%	-47%	121%	-4%	147%	-7%	-24%	-63%	85%	78%	27%	-34%	41%	-40%	16%	-51%
CG 200/350/400	54%	-21%	222%	11%	148%	-6%	200%	-32%	24%	-50%	216%	38%	106%	-18%	166%	-29%	125%	-38%
MB PPP / KGST TGT	1,62%	1,24%	1,35%	1,31%	2,53%	1,29%	1,66%	1,21%	1,40%	1,07%	1,70%	1,15%	1,62%	1,15%	1,74%	1,28%	1,40%	1,24%
MB KGST ACT / BKI	0,80%	1,62%	0,46%	1,19%	0,65%	1,72%	0,58%	2,39%	0,79%	1,97%	0,57%	1,19%	0,67%	1,68%	0,63%	2,32%	0,63%	2,24%
Residual value	21%	0%	41%	3%	20%	4%	47%	-8%	14%	-1%	32%	6%	15%	-1%	32%	-5%	26%	-8%
Results UC excl. Risks	-10%	-23%	1%	-15%	-9%	-19%	20%	-10%	-9%	-20%	3%	-4%	4%	-9%	-10%	-28%	-10%	-27%
UC incl. Risks	-10%	-22%	-2%	-15%	-9%	-19%	17%	-10%	-10%	-20%	1%	-4%	3%	-9%	-12%	-28%	-12%	-27%
LCC excl. Risks	-23%	-47%	5%	-53%	-22%	-41%	56%	-19%	-24%	-41%	9%	-10%	14%	-26%	-28%	-56%	-26%	-52%
LCC incl. Risks	-40%	-47%	-64%	-57%	-43%	-47%	-8%	-13%	-35%	-39%	-28%	-18%	-21%	-27%	-55%	-58%	-46%	-49%
LCC incl. Risks+VAT	-43%	-50%	-66%	-60%	-45%	-49%	-11%	-15%	-37%	-42%	-29%	-19%	-26%	-32%	-55%	-58%	-47%	-50%

Sensitivity analyse indexing residual value + maintenance with 2% p.a.

	1 KGSt	1 BKI	2 KGSt	2 BKI	3 KGSt	3 BKI	4 KGSt	4 BKI	5 KGSt	5 BKI	6 KGSt	6 BKI	7 KGSt	7 BKI	8 KGSt	8 BKI	9 KGSt	9 BKI
UC excl. Risks	-10%	-23%	1%	-14%	-10%	-20%	18%	-10%	-10%	-20%	2%	-5%	3%	-9%	-11%	-29%	-11%	-27%
UC incl. Risks	-10%	-23%	-1%	-14%	-10%	-19%	16%	-10%	-11%	-20%	1%	-5%	2%	-9%	-13%	-29%	-12%	-27%
LCC excl. Risks	-19%	-40%	2%	-35%	-20%	-37%	37%	-16%	-21%	-36%	4%	-10%	9%	-21%	-23%	-48%	-21%	-45%
LCC incl. Risks	-31%	-40%	-41%	-38%	-35%	-40%	-1%	-11%	-29%	-35%	-20%	-15%	-14%	-21%	-43%	-49%	-36%	-42%
LCC incl. Risks+VAT	-33%	-42%	-43%	-40%	-37%	-42%	-4%	-13%	-31%	-37%	-21%	-16%	-18%	-25%	-43%	-49%	-37%	-42%

Sensitivity analyse indexing residual value + maintenance with 1,5% p.a.

	1 KGSt	1 BKI	2 KGSt	2 BKI	3 KGSt	3 BKI	4 KGSt	4 BKI	5 KGSt	5 BKI	6 KGSt	6 BKI	7 KGSt	7 BKI	8 KGSt	8 BKI	9 KGSt	9 BKI
UC excl. Risks	-10%	-24%	1%	-14%	-10%	-21%	18%	-9%	-10%	-21%	1%	-6%	3%	-9%	-12%	-29%	-11%	-27%
UC incl. Risks	-10%	-23%	-1%	-14%	-10%	-20%	16%	-9%	-11%	-21%	0%	-6%	2%	-9%	-13%	-29%	-12%	-27%
LCC excl. Risks	-17%	-37%	2%	-30%	-19%	-35%	19%	-15%	-19%	-34%	3%	-10%	8%	-19%	-21%	-45%	-20%	-42%
LCC incl. Risks	-27%	-37%	-33%	-32%	-32%	-38%	1%	-11%	-26%	-33%	-17%	-14%	-11%	-19%	-38%	-45%	-33%	-39%
LCC incl. Risks+VAT	-30%	-39%	-35%	-34%	-34%	-39%	-1%	-13%	-29%	-35%	-18%	-15%	-15%	-23%	-38%	-46%	-33%	-40%

Key: DIN 276 construction costs: KG 200-700; DIN 18960 KG 100: capital costs; KG 200: property management costs; KG 300: operating costs; KG 310: supply (water, heat, electricity); KG 320: waste disposal; KG 330/340: cleaning; KG 350: inspection&maintenance; KG 370: taxes, insurance; KG 390: other operating costs; KG 400: repairs; KG 200 (pro rata)/KG 350/KG 400: maintenance; MB: Maintenance budget as a percentage of restoration costs p.a.; UC: Usage costs, LCC: Life cycle costs; VAT: LCC incl. risks and additional VAT revenue

European comparative study on the economics of public-private partnership life cycle costs (Germany)

PPP project no: / PPP comparison with	10 KGSt	10 BKI	11 KGSt	11 BKI	12 KGSt	12 BKI	13 KGSt	13 BKI	14 KGSt	14 BKI	15 KGSt	15 BKI	16 KGSt	16 BKI	17 KGSt	17 BKI	18 KGSt	18 BKI
DIN 276 Construction costs		-8%		-12%		-20%		-12%		-4%		-11%		-12%		-14%		-9%
DIN 277 Construction time		-19%		-26%		-38%		-35%		0% 600%		-35%		-27%		-23%	-35%	-35%
DIN 18960 CG 100		-3%		-11%		-18%		-11%		-3%		-7%		-12%		-12%		-4%
CG 100			IZZB	-12%							IZZB	-8%						
CG 200	-18%	-4%	-15%	-1%	4%	-3%	16%	12%	59%	84%	-73%	-62%	13%	41%	95%	106%	-59%	-53%
CG 300	-8%	-32%	34%	-4%	17%	2%	-15%	-62%	-19%	-45%	16%	-15%	24%	-35%	21%	-16%	46%	5%
CG 310	-24%	-46%	-24%	-44%	7%	3%	0%	-14%	-14%	-40%	-20%	4%	-50%	-55%	-33%	-33%	55%	14%
CG 320	0%	0%	0%	0%	0%	0%	0%	0%	0%	0%	0%	0%	0%	0%	0%	0%	0%	0%
CG 330/40	-34%	-26%	0%	12%	9%	56%	-7%	14%	-13%	-11%	-11%	-27%	15%	27%	-13%	-18%	-18%	-10%
CG 350	386%	-40%	591%	-14%									1097%	-58%				
CG 370	10%	-73%									197%	2%			80%	-49%	342%	5%
CG 390	-37%	56%	1398%	5226%	252%	100%					1641%				11681%	2560%	2938%	9349%
CG 400	139%	-13%	44%	-25%	171%	156%	291%	360%	82%	-6%	70%	-40%	-27%	-61%	84%	-22%	55%	-20%
CG 200/350/400	199%	-1%	117%	-16%	197%	49%	311%	17%	117%	-6%	59%	-56%	102%	-40%	100%	-40%	59%	-40%
MB PPP / KGST TGT	2%	1%	1,56%	1,17%	1%	1%	1%	1%	1,64%	1,03%	1%	1%	1,18%	1,20%	2%	1%	1,53%	1,17%
MB KGST ACT / BKI	1%	2%	0,64%	1,67%	0%	1%	0%	1%	0,55%	1,51%	1%	2%	0,65%	2,04%	1%	2%	0,61%	1,63%
Residual value	28%	-2%	27%	-1%	34%	12%	46%	0%	29%	1%	15%	-7%	28%	-12%	31%	1%	28%	-1%
Results UC excl. Risks	4%	-10%	12%	-11%	-5%	-7%	2%	-8%	6%	-8%	-2%	-16%	-5%	-24%	13%	-8%	10%	-7%
UC incl. Risks	3%	-10%	10%	-11%	-7%	-7%	1%	-8%	5%	-8%	-3%	-16%	-7%	-24%	11%	-8%	9%	-7%
LCC excl. Risks	9%	-20%	25%	-18%	-14%	-20%	8%	-22%	15%	-17%	-4%	-32%	-14%	-52%	44%	-18%	33%	-16%
LCC incl. Risks	-18%	-20%	-2%	-18%	-47%	-35%	-39%	-22%	-17%	-19%	-22%	-27%	-43%	-43%	-18%	-23%	-17%	-16%
LCC incl. Risks+VAT	-20%	-21%	-5%	-21%	-48%	-36%	-40%	-24%	-19%	-21%	-23%	-27%	-45%	-45%	-24%	-28%	-18%	-17%

Sensitivity analyse indexing residual value + maintenance with 2% p.a.	10 KGSt	10 BKI	11 KGSt	11 BKI	12 KGSt	12 BKI	13 KGSt	13 BKI	14 KGSt	14 BKI	15 KGSt	15 BKI	16 KGSt	16 BKI	17 KGSt	17 BKI	18 KGSt	18 BKI
UC excl. Risks	3%	-11%	11%	-11%	-6%	-8%	1%	-10%	5%	-9%	-3%	-16%	-4%	-23%	11%	-9%	9%	-7%
UC incl. Risks	1%	-11%	9%	-11%	-7%	-8%	-1%	-10%	4%	-9%	-4%	-16%	-6%	-23%	9%	-9%	7%	-7%
LCC excl. Risks	5%	-17%	19%	-17%	-13%	-17%	2%	-20%	10%	-15%	-6%	-28%	-9%	-40%	25%	-16%	20%	-14%
LCC incl. Risks	-14%	-17%	-1%	-17%	-37%	-28%	-30%	-21%	-12%	-16%	-18%	-24%	-28%	-34%	-9%	-19%	-10%	-14%
LCC incl. Risks+VAT	-15%	-18%	-3%	-20%	-38%	-29%	-31%	-22%	-14%	-19%	-18%	-24%	-30%	-35%	-14%	-23%	-11%	-15%
Sensitivity analyse indexing residual value + maintenance with 1,5% p.a.																		
UC excl. Risks	2%	-11%	10%	-12%	-7%	-9%	0%	-11%	5%	-9%	-4%	-17%	-4%	-23%	10%	-9%	8%	-8%
UC incl. Risks	1%	-11%	8%	-12%	-8%	-9%	-1%	-11%	3%	-9%	-4%	-17%	-5%	-23%	9%	-9%	7%	-8%
LCC excl. Risks	3%	-16%	16%	-17%	-13%	-17%	0%	-20%	8%	-15%	-6%	-26%	-7%	-36%	20%	-16%	17%	-14%
LCC incl. Risks	-12%	-16%	0%	-17%	-33%	-25%	-27%	-20%	-10%	-16%	-16%	-23%	-23%	-31%	-6%	-18%	-7%	-14%
LCC incl. Risks+VAT	-13%	-17%	-2%	-19%	-34%	-26%	-28%	-21%	-12%	-18%	-16%	-23%	-25%	-32%	-10%	-21%	-8%	-14%

Key: DIN 276 construction costs: KG 200-700; DIN 18960 KG 100: capital costs; KG 200: property management costs; KG 300: operating costs; KG 310: supply (water, heat, electricity); KG 320: waste disposal; KG 330/340: cleaning; KG 350: inspection&maintenance; KG 370: taxes, insurance; KG 390: other operating costs; KG 400: repairs; KG 200 (pro rata)/KG 350/KG 400: maintenance; MB: Maintenance budget as a percentage of restoration costs p.a.; UC: Usage costs, LCC: Life cycle costs; VAT: LCC incl. risks and additional VAT revenue

ANNEX A2   Result of qualitative comparison of 18 PPP projects / KGST / BKI

European Comparative PPP Study (Germany)
Qualitative evaluation of PPP construction and consulting services

PPP project no:	1			2			3			4			5			6			7			8			9		
	PPP	KGST	BKI	PPP	KGST	BKI	PPP	KGST	BKI	PPP	KGST	BKI	PPP	KGST	BKI	PPP	KGST	BKI	PPP	KGST	BKI	PPP	KGST	BKI	PPP	KGST	BKI
Investment phase	1,7	3	3	1,2	1,8	1,8	2,4	3,0	3,0	2,2	2,8	2,8	2,6	3,0	3,0	2,1	2,7	2,7	2,4	3,0	3,0	2,4	3,0	3,0	2,4	3,0	3,0
Building standard	2	3	3	1,3	1,3	1,3	3	3	3	2,8	2,8	2,8	3	3	3	2,5	2,5	2,5	3	3	3	3	3	3	3	3	3
Cost certainty	1	3	3	1	3	3	1	3	3				3	3	3	1	3	3	1	3	3	1	3	3	1	3	3
Construction time	1	3	3	1	3	3	1	3	3	1	3	3	1	3	3	1	3	3	1	3	3	1	3	3	1	3	3
Interest rate hedging	1	3	3	1	3	3	1	3	3	1	3	3	1	3	3	1	3	3	1	3	3	1	3	3	1	3	3
Usage phase	1,9	3,4	2,3	2,5	3,3	2,9	2,1	3,3	2,6	2,3	3,3	2,6	2,3	3,3	2,5	2,4	3,3	2,7	2,4	3,3	2,8	2,6	3,3	2,6	2,7	3,4	2,7
Financing	3	3	3	3	3	3	2	3	3	3	3	3	2	3	3	3	3	3	3	3	3	3	3	3	3	3	3
Property management	1	3	1	2	3	3	1	3	2	1	3	2	1	3	3	1	3	1	1	3	3	3	3	3	3	3	3
Supply	1	4	5	2	3	3	2	3	3	2	3	4	3	3	3	2	3	4	2	3	4	2	3	3	2	4	4
Waste disposal	3	3	2	3	3	3	3	3	3	3	3	3	3	3	3	3	3	3	3	3	3	3	3	3	3	3	3
Cleaning	2	3	2	3	3	3	3	3	3	3	3	3	3	3	3	3	3	3	3	3	3	3	3	3	3	3	3
Inspection and maintenance	1	5	1	1	5	1,5	1	5	1	1	5	1	1	5	1	2	5	1	1	5	1	1	5	1	1	5	1
Taxes and assurance	2	3	1	3	3	3	3	3	3	3	3	3	2	3	2	3	3	3	3	3	3	3	3	3	3	3	3
Cafeteria operation	3	3	3				3	3	3	3	3	3	3	3	3	3	3	3	3	3	3	3	3	3	3	3	3
Ohter operating costs	1	3	3	3	3	3,3	2	3	3	2	3	3	3	3	3	2	3	3	3	3	3	3	3	3	3	3	3
Repairs / maintenance	2	4	2	2,8	4	3	1	4	1	2	4	1	2	4	1	2	4	3	2	4	2	2	4	1	2,5	4	1
Residual value	2,1	3,9	2,1	2,0	4,0	3,0	1,0	3,5	2,0	2,0	4,0	1,0	2,0	4,0	1,0	1,5	4,0	3,0	2,0	4,0	2,0	2,0	4,0	1,0	2,5	4,0	1,0
Total result	1,9	3,4	2,5	1,9	3,0	2,5	1,8	3,3	2,5	2,2	3,4	2,1	2,3	3,4	2,2	2,0	3,3	2,8	2,3	3,4	2,6	2,3	3,4	2,2	2,5	3,5	2,2

[3] Quality assessment: 1 = very good / 2 = good / 3 = satisfactory / 4 = sufficient / 5 = unsatisfactory

No PPP contract component

European Comparative PPP Study (Germany)

Qualitative evaluation of PPP construction and consulting services

PPP project no:	10			11			12			13			14			15			16			17			18			Ø		
	PPP	KGST	BKI	PPP	KGST	BKI	PPP	KGST	BKI	PPP	KGST	BKI	PPP	KGST	BKI	PPP	KGST	BKI	PPP	KGST	BKI	PPP	KGST	BKI	PPP	KGST	BKI	PPP	KGST	BKI
Investment phase	2,4	3,0	3,0	2,4	3,0	3,0	1,7	2,3	2,3	1,4	2,0	2,0	2,4	3,0	3,0	2,4	3,0	3,0	2,3	2,8	2,8	2,4	3,0	3,0	2,4	3,0	3,0	2,2	2,8	2,8
Building standard	3	3	3	3	3	3	2	2	2	1,5	1,5	1,5	3	3	3	3	3	3	2,8	2,8	2,8	3	3	3	3	3	3	2,7	2,7	2,7
Cost certainty	1	3	3	1	3	3	1	3	3	1	3	3	1	3	3				2	3	3	1	3	3				1,2	3,0	3,0
Construction time	1	3	3	1	3	3	1	3	3	1	3	3	1	3	3	1	3	3	1	3	3	1	3	3	1	3	3	1,0	3,0	3,0
Interest rate hedging	1	3	3	1	3	3	1	3	3	1	3	3	1	3	3	1	3	3	1	3	3	1	3	3	1	3	3	1,0	3,0	3,0
Usage phase	2,3	3,3	2,8	2,6	3,3	2,8	2,6	3,4	2,9	2,6	3,4	2,8	2,5	3,3	2,8	2,6	3,3	2,6	2,7	3,3	2,7	2,3	3,3	2,6	2,7	3,3	2,8	2,5	3,3	2,7
Financing	2	3	3	3	3	3	3	3	3	3	3	3	3	3	3	2	3	3	3	3	3	3	3	3	2	3	3	2,7	3,0	3,0
Property management	2	3	2,5	3	3	3	2	3	3	2	3	3	2	3	3	2	3	3	2	3	3	2	3	3	2	3	3	1,8	3,0	2,7
Supply	2	3	4	2	3	4	3	3	3	3	3	3	2	3	4	2,5	3	2,25	2	3	3,5	2	3	3	4	3	4	2,3	3,1	3,5
Waste disposal	3	3	3	3	3	3	3	3	3	3	3	3	3	3	3	3	3	3	3	3	3	3	3	3	3	3	3	3,0	3,0	2,9
Cleaning	3	3	3	3	3	3	3	3	3	3	3	3	3	3	3	3	3	3	3	3	3	3	3	3	3	3	3	2,9	3,0	2,9
Inspection and maintenance	1	5	1	1	5	1	1	5	1	1	5	1	1	5	1	2,5	5	1	2	5	1	2	5	1	2,5	5	1	1,4	5,3	1,1
Taxes and assurance	2	3	3	3	3	3	3	3	3	3	3	3	3	3	3	3	3	3	3	3	3	3	3	3	3	3	3	2,8	3,0	2,8
Cafeteria operation	3	3	3	3	3	3							3	3	3	3	3	3	3	3	3				3	3	3	3	3	3
Ohter operating costs	3	3	3	3	3	3	3	3	3	3	3	3	3	3	3	2	3	3	3	3	3	2	3	3	2	3	3	2,6	3,0	3,0
Repairs / maintenance	2,25	4	2	2	4	1,5	2,5	4,5	3,75	2,5	4,5	3	2	4	2,25	2,5	4	1,5	3	4	1	1	4	1,5	2,5	4	1,5	2,1	4,1	1,8
Residual value	2,3	4,0	2,0	2,3	4,0	2,0	2,5	4,5	3,8	3,0	4,5	3,0	2,0	4,0	2,3	2,5	4,0	1,5	3,0	4,0	1,0	1,0	4,0	1,5	2,5	4,0	2,0	2,1	4,0	2,0
Total result	2,3	3,4	2,6	2,4	3,4	2,6	2,3	3,4	3,0	2,3	3,3	2,6	2,3	3,4	2,7	2,5	3,4	2,4	2,7	3,4	2,2	1,9	3,4	2,4	2,5	3,4	2,6	2,2	3,4	2,5

[3] Quality assessment: 1 = very good / 2 = good / 3 = satisfactory / 4 = sufficient / 5 = unsatisfactory

▢ No PPP contract component

A3 Overview of results — Average values median, arithmetic mean, gross floor area-weighted mean, average of mean values

European comparative PPP study (GER) Results with different mean values*		NEW BUILD Projects no 1-16 MEDIAN		NEW BUILD Projects no 1-16 AVERAGE		NEW BUILD Projects no 1-16 AV WEIGHTED		NEW BUILD Projects no 1-16 Ø MEAN VALUES		RENOVATION Projects no 17+18 AV WEIGHTED		NEW+RENOVATION Projects no 1-18 MEDIAN		NEW+RENOVATION Projects no 1-18 AVERAGE		NEW+RENOVATION Projects no 1-18 AV WEIGHTED		NEW+RENOVATION Projects no 1-18 Ø MEAN VALUES	
		PPP/KGST	PPP/BKI	PPP/KGST	PPP/BKI	PPP/KGST	PPP/BKI	PPP/KGST	PPP/BKI	PPP/KGST	PPP/BKI	PPP/KGST	PPP/BKI	PPP/KGST	PPP/BKI	PPP/KGST	PPP/BKI	PPP/KGST	PPP/BKI
DIN 276 CG 200-700	Construction costs (PPP/BKI)	-15%	-15%	-16%	-16%	-20%	-20%	-17%	-17%	-11%	-11%	-13%	-13%	-16%	-16%	-19%	-19%	-16%	-16%
DIN 277	Construction time (PPP/BKI)	-35%	-35%	-35%	-35%	-27%	-27%	-30%	-30%	-29%	-29%	-35%	-35%	-30%	-30%	-27%	-27%	-30%	-30%
DIN 18960 CG 100	Capital costs	-15%	-15%	-14%	-14%	-17%	-17%	-15%	-15%	-8%	-8%	-12%	-12%	-13%	-13%	-16%	-16%	-14%	-14%
CG 200	Property management	11%	18%	11%	23%	15%	27%	12%	23%	18%	27%	11%	18%	12%	23%	18%	30%	14%	24%
CG 300	Operating costs	18%	-29%	21%	-21%	18%	-21%	19%	-24%	33%	-5%	19%	-21%	22%	-19%	18%	-20%	20%	-20%
CG 310	Supply	-22%	-39%	-18%	-30%	-11%	-27%	-17%	-32%	11%	-10%	-22%	-37%	-15%	-28%	-11%	-27%	-16%	-31%
CG 320	Waste disposal	0%	0%	0%	0%	0%	0%	0%	0%	0%	0%	0%	0%	0%	0%	0%	0%	0%	0%
CG 330/340	Cleaning	6%	12%	7%	13%	2%	2%	5%	9%	-15%	-14%	1%	9%	4%	10%	1%	0%	2%	6%
CG 350	Inspection and maintenance	655%	-27%	747%	-19%	653%	-17%	685%	-21%			656%	-27%	763%	-19%	653%	-17%	691%	-21%
CG 360	Energy management																		
CG 370	Taxes, insurances	178%	-17%	209%	20%	149%	-24%	179%	-7%	211%	-22%	178%	-17%	210%	9%	148%	-25%	179%	-11%
CG 390	Other operating costs	317%	90%	690%	3289%	359%	1600%	455%	1660%	7310%	5955%	1210%	1330%	1894%	3822%	1545%	1957%	1549%	2370%
CG 400	Repairs	57%	-30%	74%	10%	67%	-9%	66%	-10%	69%	-21%	62%	-23%	74%	6%	68%	-10%	68%	-9%
CG 200/350/400	Maintenance budget	137%	-17%	148%	-12%	126%	-14%	137%	-14%	80%	-40%	121%	-19%	140%	-15%	123%	-16%	128%	-17%
	in % restoration costs p.a. PPP and KGST-Target	1,6%	1,2%	1,6%	1,2%	1,8%	1,2%	1,6%	1,2%	1,8%	1,2%	1,6%	1,2%	1,6%	1,2%	1,8%	1,2%	1,6%	1,2%
	in % restoration costs p.a. KGST-Actual and BKI	0,6%	1,7%	0,6%	1,7%	0,6%	1,7%	0,6%	1,7%	0,6%	1,7%	0,6%	1,7%	0,6%	1,7%	0,6%	1,7%	0,6%	1,7%
	Restwert	28%	-1%	29%	-1%	25%	0%	27%	0%	30%	0%	28%	-1%	29%	-1%	25%	0%	27%	0%
Usage costs excluding risk costs		0%	-13%	-1%	-15%	-4%	-17%	-1%	-15%	11%	-7%	2%	-11%	1%	-14%	-2%	-16%	0%	-14%
Usage costgs including risk costs		-2%	-13%	-2%	-15%	-5%	-16%	-3%	-15%	10%	-7%	-1%	-11%	-1%	-14%	-3%	-16%	-2%	-13%
Lefe cycle costs excluding risk costs		1%	-29%	-1%	-33%	-9%	-35%	-3%	-32%	39%	-17%	7%	-24%	3%	-34%	-5%	-34%	2%	-30%
Life cycle costs including risk costs		-37%	-31%	-33%	-34%	-34%	-37%	-35%	-34%	-17%	-20%	-31%	-27%	-31%	-32%	-33%	-36%	-32%	-32%
Life cycle costs including risk costs and VAT revenue		-39%	-34%	-35%	-36%	-36%	-39%	-37%	-36%	-21%	-23%	-33%	-30%	-34%	-34%	-35%	-38%	-34%	-34%

* Mean values: MEDIAN, arithmetic AVERAGE, average value weighted by GFA (AV WEIGHTED), average of the mean values (Ø MEAN VALUES)

A.3 Overview of results of the individual 18 PPP projects

	Economic efficiency study — PPP project no 1 (greenfield / education sector)	Index	PPP[1]	KGSt1)	BKi[1]	PPP - KGSt	PPP - BKI	PPP/ KGSt	PPP/ BKI	Quality[3] PPP	KGSt	BKI	Value	Remarks	QCO[4]
DIN 276	Construction costs									1,7	3,0	3,0	100%	A. Construction quality, cost and schedule efficiency	
DIN 277	GFA in m²		57.649	57.649	57.649										
	Construction time in months		19	30	30	-11	-11	-36%	-36%	1	3	3	10%	Schedule reliability 100% PPP/BKI construction costs p.m. factor 9.5	2
DIN 276	Construction costs excl. CG 760		72.704.711 €	92.030.314 €	92.030.314 €	-19.325.602 €	-19.325.602 €	-21%	-21%	2	3	3	70%	PPP: High building standard, BKI: Medium building standard	
	Construction costs /sqm GFA in T€		1.261 €	1.596 €	1.596 €	-335 €	-335 €	-21%	-21%	1	3	3	10%	PPP: 100% cost certainty	
CG 760	Interim Financing		1.320.793 €	1.706.447 €	1.706.447 €	-385.665 €	-385.665 €	-23%	-23%						
	Interest rate		2.698%	1.938%	1.938%	0.76%	0.76%	39%	39%	1	3	3	10%	Interest rate hedging promotes high cost and deadline security	
DIN 276	Construction costs incl. CG 760		74.025.504 €	93.736.761 €	93.736.761 €	-19.711.257 €	-19.711.257 €	-21%	-21%						1
	Construction costs /sqm GFA in T€		1.284 €	1.626 €	1.626 €	-342 €	-342 €	-21%	-21%						
DIN 18960	Usage Costs excluding risk costs		207.536.127 €	229.461.266 €	270.628.235 €	-21.925.138 €	-63.092.108 €	-10%	-23%	1,9	3,4	2,3	100%	B. High-quality maintenance/energy/property management	
	Usage costs including risk costs*		209.811.436 €	232.167.230 €	270.628.235 €	-22.355.794 €	-60.816.799 €	-10%	-22%						
CG 100	Capital costs		128.830.168 €	158.404.112 €	158.404.112 €	-29.573.945 €	-29.573.945 €	-19%	-19%					PPP: Forfaiting with waiver of defence	1
CG 100	Interest		54.626.164 €	64.667.351 €	64.667.351 €	-10.041.188 €	-10.041.188 €	-16%	-16%	3	3	3	10%		
CG 100	Interest rates		4.899%	4.63%	4.63%	0.27%	0.27%	6%	6%						
CG 200	Property management	1.42%	19.140.218 €	14.111.330 €	14.466.021 €	5.028.888 €	4.674.197 €	36%	32%	1	3	1	10%		1
CG 210	Personnel costs	1.42%	18.283.449 €	14.111.330 €											
CG 210	Own costs of the city	1.42%	856.770 €												
CG 300	Operating costs		44.156.659 €	37.730.973 €	64.735.971 €	6.425.686 €	-20.579.313 €	18%	-32%						
CG 310	Supply costs[2]		7.747.870 €	19.737.109 €	24.162.986 €	-11.989.239 €	-16.415.116 €	-61%	-68%	1	4	5	10%	PPP - Heating+electricity: 58% < EnEV 2011 / 42% < VDI / 66% < KGSt / 73% < BKI	1
CG 311	Water	1.49%	303.718 €	339.921 €	489.859 €	-36.203 €	-186.141 €	-11%	-38%						
CG 312	Heating	1.12%		5.153.020 €	7.561.597 €	-5.153.020 €	-7.561.597 €								
CG 313	Electricity	3.57%	7.444.152 €	6.177.280 €	8.044.641 €	1.266.872 €	-600.490 €	21%	-7%						
CG 320	Wasrte disposal	1.42%	3.232.204 €	3.232.204 €	3.232.204 €	0 €	0 €	0%	0%	3	3	2	10%		1
CG 330	Cleaning	2.49%	16.126.763 €	10.795.191 €	16.690.529 €	5.331.572 €	-563.766 €	49%	-3%	2	3	2	10%	PPP service catalogue > DIN 77400, cleaning interval 31% > KGSt	2
CG 351	Operation of technical systems - Energy management	2.20%	299.309 €		5.643.837 €	299.309 €	-5.344.528 €		-95%						
CG 350	Inspection and maintenance	2.20%	12.650.028 €	1.669.115 €	10.883.011 €	10.980.913 €	1.767.018 €	658%	16%	1	5	1	10%	PPP: 216 maintenance measures from 29 companies (2019) due to operator liability	
CG 370	Taxes and insurance	1.56%	900.481 €	890.691 €	2.839.123 €	9.790 €	-1.938.642 €	1%	-68%	2	3	1	10%	PPP: 1.9 Mio € operating guarantee	
CG 391	Cafeteria operation	1.42%	924.694 €	924.694 €	924.694 €	0 €	0 €	0%	0%	3	3	3	10%		3
CG 392	Guarantee commission'	1.42%	2.275.309 €							1				Guarantee commission serves quality assurance	
CG 393	Other operating costs	1.42%		481.968 €	359.587 €						3	3	10%		1
CG 400	Repair costs	2.92%	17.684.392 €	21.920.815 €	33.022.131 €	-4.236.423 €	-15.337.739 €	-20%	-46%						
CG 400	Repair costs	2.92%	17.684.392 €	19.214.850 €	33.022.131 €	-1.530.459 €	-15.337.739 €	-8%	-46%						
CG 460	Risk of structural damage (maintenance)'	2.92%		2.705.964 €		-2.705.964 €	2.705.964 €							CPP: Risk of structural damage due to insufficient maintenance budgets	
CG 200/350/400	Maintenance		40.193.513 €	26.041.287 €	50.892.057 €	14.152.225 €	-10.698.545 €	54%	-21%	2	4	2	10%		1
CG 200/350/400	in % of restoration costs p.a. / CGST-Target	1.24%	1.62%	0.80%	1.62%										
	Revenue / residual value excluding risk costs		135.304.090 €	135.304.090 €	135.304.090 €	0 €	0 €	0%	0%						
	Revenue / residual value including risk costs		141.799.977 €	118.330.966 €	141.799.977 €	23.469.011 €	0	20%	0%	2,1	3,9	2,1	100%	C. Chance of good residual value without maintenance backlog	
	Revenue / residual value including risk costs and VAT		144.903.767 €	118.330.966 €	141.799.977 €	26.572.801 €	3.103.790	22%	2%						
CG 500	Car park management	1.42%	6.685.526 €	6.685.526 €	6.685.526 €	0 €	0 €	0%	0%	3	3	3	5%		
CG 500	Residual value	2.92%	135.114.451 €	111.645.440 €	135.114.451 €	23.469.011	0	21%	0%	2	4	2	95%	(1) Relationship between maintenance budget / useful life	1
	Useful life		90	62	90									(cf. PPP School Study2019, margin no. 137) / (2) PPP/BKI: more favourable	
	Residual value risk (maintenance)'		6.495.887 €	-16.973.124 €	6.495.887 €	23.469.011	0	-138%	0%					construction costs lead to hidden reserves (3) Indexing residual value	
CG 500	Additional VAT revenue federal/state/local authorities		3.103.790 €												
	Life cycle costs excluding risk costs (Nominal)		72.232.038 €	94.157.176 €	135.324.145 €	-21.925.138 €	-63.092.108 €	-23%	-47%						
	(NPV)		77.440.715 €	90.921.862 €	114.751.565 €	-13.481.147 €	-37.310.850 €	-15%	-33%						
	Life cycle costs including risk costs (Nominal)		68.011.459 €	113.836.264 €	128.828.258 €	-45.824.805 €	-60.816.799 €	-40%	-47%	1,9	3,4	2,5	(A+B+C)/3	PPP incentive structures promote quality management: cost caps, bonus/malus, savings participation, operator liability	1,3
	(NPV)		76.597.390 €	97.561.676 €	112.559.823 €	-20.964.287 €	-35.962.434 €	-21%	-32%						
	Life cacle costs incl. risk costs + additional VAT revenue (Nominal)		64.907.669 €	113.836.264 €	128.828.258 €	-48.928.595 €	-63.920.589 €	-43%	-50%						
	(NPV)		74.841.743 €	97.561.676 €	112.559.823 €	-22.719.933 €	-37.718.080 €	-23%	-34%						

1) Nominal values indexed over 25 years, base year 2011, net present value (NPV) - discount rate 4.699 % 2) KG 310 1 = MAX 2 = PRE Supply costs

3) Quality assessment 1 = very good / 2 = good / 3 = satisfactory / 4 = sufficient / 5 = unsatisfactory / 4) Questionnaire Construction and operating performance

Economic efficiency study — PPP project no 2 (greenfield / education sector)		Index	PPP[1]	KGSt[1]	BKI[1]	PPP - KGSt	PPP - BKI	PPP/ KGSt	PPP/ BKI	Quality[2] PPP	KGSt	BKI	Value	Remarks	QCO[4]
DIN 276	**Construction costs**									**1,2**	**1,8**	**1,8**	**100%**	**A. Construction quality, cost and schedule efficiency**	**2**
DIN 277	GFA in m²		4.298	4.298	4.298										
	Construction time in months		17	26	26	-9	-9	-35%	-35%	1	3	3	10%	Schedule reliability: completion 4 5% earlier	
DIN 276	Construction costs excl. CG 760		10.825.913 €	13.191.612 €	13.191.612 €	-2.365.699	-2.365.699	-18%	-18%	1,25	1,25	1,25	70%	High building standard	2
	Construction costs /sqm GFA in T€		2.519 €	3.069 €	3.069 €	-550	-550	-18%	-18%	1	3	3	10%	PPP: High cost certainty: Actual cost 1% < Target costs	
CG 760	Interim Financing		108.614 €	148.518 €	148.518 €	-39.904	-39.904	-27%	-27%	1	3	3	10%	Interest rate hedging promotes high cost and deadline security	
	Interest rate		1.80%	1.20%	1.20%	0,60%	0,60%	50%	50%						
DIN 276	Construction costs incl. CG 760		10.934.527 €	13.340.130 €	13.340.130 €	-2.405.603	-2.405.603	-18%	-18%						
	Construction costs /sqm GFA in T€		2.544 €	3.104 €	3.104 €	-560	-560	-18%	-18%						
DIN 18960	**Usage Costs excluding risk costs**		**21.675.279**	**21.504.197**	**25.539.247**	**171.082**	**-3.863.968**	**1%**	**-15%**	**2,5**	**3,3**	**2,9**	**100%**	**B. High-quality maintenance/energy/property management**	**2**
	Usage costs including risk costs*		**21.675.279**	**22.036.686**	**25.539.247**	**-361.407**	**-3.863.968**	**-2%**	**-15%**						
CG 100	Capital costs		12.383.346 €	14.902.107 €	14.902.107 €	-2.518.761	-2.518.761	-17%	-17%	3	3	3	12,5%	PPP: Forfaiting with waiver of defence	
CG 100	Interest		1.448.819 €	1.561.977 €	1.561.977 €	-113.158	-113.158	-7%	-7%						
CG 100	Interest rates		1,05%	0,90%	0,90%	0,15%	0,15%	17%	17%						2
CG 200	Property management	1,42%	1.420.852 €	1.209.366 €	1.149.806 €	211.486	271.046	17%	24%	2	3	3	12,5%	VAT (Personnel) / CAFM / Quality assurance/ Remuneration of personnel	
CG 210	Personnel costs	1,42%	1.356.921 €												
CG 210	Own costs of the city	1,42%	63.930 €												
CG 300	Operating costs		5.770.914 €	3.713.201 €	5.549.097 €	2.057.713 €	221.817 €	55%	4%						
CG 310	Supply costs 2)		552.031 €	1.464.529 €	1.479.659 €	912.498 €	927.627 €	-62%	-63%	2	3	3	12,5%	Cost advantage speaks in favour of lower energy consumption	2
CG 311	Water	1,49%	37.211 €	203.073 €	202.059 €	-165.862	-164.848	-82%	-82%					as yet no benchmark comparison of consumption quantities	
CG 312	Heating	1,12%	110.248 €	662.513 €	653.859 €	-552.265 €	-543.611 €	-83%	-83%						
CG 313	Electricity	3,57%	404.572 €	598.943 €	623.741 €	-194.371	-219.169	-32%	-35%						
CG 320	Wasrte disposal	1,42%	224.112 €	224.112 €	224.112 €	0	0	0%	0%	3	3	3	12,5%	No qualitative study to date	
CG 330	Cleaning	2,49%	2.381.634 €	1.789.367 €	1.892.276 €	592.267	489.357	33%	26%	3	3	3	12,5%	No qualitative study to date	3
CG 351	Operation of technical systems - Energy management	2,20%	52.056 €												
CG 352	Inspection and maintenance	2,20%	2.561.081 €	145.288 €	1.365.080 €	2.415.792	1.196.001	1663%	88%	1	5	1,5	12,5%	PPP incentive structure: service levels and operator liability force	
CG 370	Taxes and insurance	1,56%	0 €	77.642 €	292.389 €	-77.642	-292.389			3	3	3	12,5%	high-quality maintenance & inspection	
CG 393	Other operating costs	1,42%	0 €	12.263 €	295.581 €	-12.263	-295.581			3	3	3.0			
CG 400	Repair costs	2,92%	2.100.167 €	1.679.523 €	3.938.237 €	420.644 €	1.838.070 €	25%	-47%					Good maintenance management: Service levels, user requests for	
CG 460	Risk of structural damage (maintenance)*	2,92%		532.490 €		-532.490	532.490							compliance with response/repair times: Cost ceilings: component-specific	
CG 200/350/400 Maintenance		2,92%	5.870.613 €	1.824.811 €	5.303.317 €	4.045.802	567.296	222%	11%	2,75	4	3.25	12,5%	maintenance calculation eduction in charges for poor performance.	2
CG 200/350/400 in % of restoration costs p.a. / CGST-Target	1,31%		1,35%	0,46%	1,19%									reserve account; CPP: damage risks due to low maintenance budgets	
Revenue / residual value excluding risk costs			**18.304.328 €**	**18.304.328 €**	**18.304.328 €**	**0 €**	**0**	**0%**	**0%**						
Revenue / residual value including risk costs			**18.407.046 €**	**13.040.560 €**	**17.866.425 €**	**5.366.486 €**	**540.620**	**41%**	**3%**	**2,0**	**4,0**	**3,0**	**100%**	**C. Chance of good residual value without maintenance backlog**	
Revenue / residual value including risk costs and VAT			**18.631.221 €**	**13.040.560 €**	**17.866.425 €**	**5.590.661 €**	**764.795**	**43%**	**4%**						
CG 500	Residual value including maintenance risk	2,92%	18.407.046 €	13.040.560 €	17.866.425 €	5.366.486	540.620	41%	3%	2,0	4.0	3.0	100%	(1) Relationship between maintenance budget / useful life	1
	Useful life		81	49	76	32	5	65%	7%					(cf. PPP School Study2019, margin no. 137) / (2) PPP/BKI: more favourable	
	Residual value risk (maintenance)*		102.718 €	-5.263.768 €	-437.903 €	5.366.486	540.620	-102%	-123%					construction costs lead to hidden reserves (3) Indexing residual value	
CG 500	Additional VAT revenue federal/state/local authorities		224.175 €												
Life cycle costs excluding risk costs	Nominal		3.370.951 €	3.199.869 €	7.234.919 €	171.082 €	-3.863.968 €	5%	-53%						
	NPV		4.654.359 €	4.528.081 €	8.117.315 €	126.278 €	-3.462.956 €	3%	-43%						
Life cycle costs including risk costs*	Nominal		3.268.233 €	8.996.126 €	7.672.822 €	-5.727.893 €	-4.404.589 €	-64%	-57%	**1,9**	**3,0**	**2,5**	**(A+B+C)/3**	PPP incentive structures promote quality management: cost caps, bonus/malus, savings participation, operator liability	**2,0**
	NPV		4.571.515 €	9.202.845 €	8.470.489 €	-4.631.330 €	-3.898.974 €	-50%	-46%						
Life cacle costs incl. risk costs + additional VAT revenue	Nominal		3.044.058 €	8.996.126 €	7.672.822 €	-5.952.068 €	-4.628.764 €	-66%	-60%						
	NPV		4.571.315 €	9.202.845 €	8.470.489 €	-4.631.530 €	-3.899.174 €	-50%	-46%						

1) Nominal values indexed over 25 years, base year 2018, net present value (NPV) - discount rate 0.9 % 2) KG 310 1 = MAX, 2 = PRE: Supply costs
3) Quality assessment: 1 = very good / 2 = good / 3 = satisfactory / 4 = sufficient / 5 = unsatisfactory / 4) Questionnaire Construction and operating performance

Economic efficiency study — PPP project no 3 (greenfield/brownfield / education)	Index	PPP[1]	KGSt[1]	BKI[1]	PPP - KGSt	PPP - BKI	PPP/ KGSt	PPP/ BKI	Q PPP	Q KGSt	Q BKI	Value	Quality[3] — Remarks	QCO[4]
DIN 276 Construction costs									**2,4**	**3,0**	**3,0**	**100%**	**A. Construction quality, cost and schedule efficiency**	
DIN 277 GFA in m²		80.421	80.421	80.421										
Construction time in months		24	30	30	-6	-6	-19%	-19%	1	3	3	10%	Schedule reliability 100%. PPP/BKI construction costs p.m. factor 6	1
DIN 276 Construction costs excl. CG 760		95.273.903 €	140.752.531 €	140.752.531 €	-45.478.628 €	-45.478.628 €	-32%	-32%	3	3	3	70%	Medium building standard	2
Construction costs /sqm GFA in T€		1.185 €	1.750 €	1.750 €	-566 €	-566 €	-32%	-32%	1	3	3	10%	PPP Cost certainty: 99.24% (additional user request)	
CG 760 Interim Financing		3.112.662 €	4.458.912 €	4.458.912 €	-1.346.250 €	-1.346.250 €	-30%	-30%	1	3	3	10%	Interest rate hedging promotes high cost and deadline security	
Interest rate		3.880%	3.280%	3.280%	0.60%	0.60%	18%	18%						
DIN 276 Construction costs incl. CG 760		98.386.565 €	145.211.443 €	145.211.443 €	-46.824.878 €	-46.824.878 €	-32%	-32%						
Construction costs /sqm GFA in T€		1.223 €	1.806 €	1.806 €	-582 €	-582 €	-32%	-32%						
DIN 18960 Usage Costs excluding risk costs		**285.165.537 €**	**312.807.207 €**	**353.113.814 €**	**-27.641.671 €**	**-67.948.277 €**	**-9%**	**-19%**	**2,1**	**3,3**	**2,6**	**100%**	**B. High-quality maintenance/energy/property management**	
Usage costs including risk costs*		**287.607.671 €**	**316.100.268 €**	**353.113.814 €**	**-28.492.598 €**	**-65.506.143 €**	**-9%**	**-19%**						
CG 100 Capital costs		168.911.233 €	228.627.700 €	228.627.700 €	-59.716.466 €	-59.716.466 €	-26%	-26%					Project finance (PFI): higher interest rates serve quality management (QM)	1
CG 100 Interest		63.824.121 €	83.416.257 €	83.416.257 €	-19.592.136 €	-19.592.136 €	-23%	-23%	2	3	3	9.1%		
CG 100 Interest rates		5.529%	4.929%	4.929%	1%	1%	12%	12%						
CG 200 Property management	1,42%	19.101.141 €	16.968.571 €	14.646.297 €	2.132.570 €	4.454.844 €	13%	30%	1	3	3	9.1%	High-quality property management: additional costs PFI promote QM	1
CG 210 Personnel costs	1,42%	18.216.496 €											VAT (Personnel costs) / CAFM / Quality assurance/ Remuneration of personnel	
CG 210 Own costs of the city	1,42%	684.645												
CG 300 Operating costs		60.055.063 €	49.334.681 €	68.767.012 €	8.278.248 €	-11.154.083 €	17%	-16%						
CG 310 Supply costs[2]		18.190.635 €	16.060.263 €	21.946.022 €	2.130.372 € -	3.755.387 €	13%	-17%	2	3	3	9.1%	Expected 9% savings compared to max. guaranteed quantities, risk PPP company 4%	1
CG 311 Water	1,49%	543.082 €	669.352 €	946.157 € -	126.270 € -	403.076 €	-19%	-43%					Expected 13% savings compared to max. guaranteed quantities, risk PPP company 3%	
CG 312 Heating	1.12%	5.294.535 €	9.389.562 €	13.600.756 € -	4.095.027 € -	8.306.220 €	-44%	-61%					Expected 15% savings compared to max. guaranteed quantities, risk PPP company 2%	
CG 313 Electricity	3,57%	12.353.018 €	6.001.349 €	7.399.109 €	6.351.669 €	4.953.909 €	106%	67%					Voraus: 7% Einsparungen ggü. max. Garantiemengen	
CG 320 Waste disposal	1,42%	2.591.257 €	2.591.257 €	2.591.257 €	- €	- €	0%	0%	3	3	3	9.1%	No qualitative study to date	2
CG 330 Cleaning	2,49%	17.740.421 €	24.306.057 €	22.079.366 € -	6.565.636 € -	4.338.945 €	-27%	-20%	3	3	3	9.1%	No qualitative study to date	2
CG 350 Inspection and maintenance	2,20%	13.550.735 €	1.863.716 €	15.214.711 €	11.687.019 € -	1.663.976 €	627%	-11%	1	5	1	9.1%	High quality maintenance management	
CG 370 Taxes and insurance	1,56%	2.406.702 €	928.497 €	3.775.093 €	1.476.204 € -	1.368.391 €	159%	-36%	3	3	3	9.1%	Mehrkosten PFI förderne QM	1
CG 390 Other operating costs	1,42%	5.575.313 €	3.584.890 €	3.160.562 €	1.990.424 €	2.414.751 €	56%	76%						
CG 391 Cafeteria operation	1,42%	2.992.827 €	2.992.827 €	2.992.827 €	- €	- €	0%	0%	3	3	3	9.1%	No qualitative study to date	2
CG 393 Letter of comfort*	1,42%	2.442.134 €											Additional costs PFI promote QM	
CG 393 Other operating costs	1,42%	140.353 €	592.063 €	167.735 € -	451.710 € -	27.383 €	-76%	-20%	2	3	3	0.0%		
CG 400 Repair costs	2,92%	39.540.233 €	17.876.256 €	41.072.805 €	21.663.977 €	-1.532.572 €	121%	-4%	1	4	1	0.0%	Good maintenance management: Service levels, user requests for	1
CG 460 Risk of structural damage (maintenance)*	2,92%		3.293.061 €		3.293.061 €	3.293.061 €							compliance with response/repair times; Cost ceilings, component-specific	
CG 200/350/400 Maintenance		56.911.197 €	22.957.828 €	60.809.118 €	33.953.369 € -	3.897.922 €	148%	-6%					maintenance calculation eduction in charges for poor performance	1
CG 200/350/400 in % of restoration costs p.a. / CGST-Target	1.29%	2.23%	0.65%	1.72%									reserve account: CPP: damage risks due to low maintenance budgets	
Revenue / residual value including risk costs and VAT		**206.157.005 €**	**168.674.979 €**	**194.560.972 €**	**37.482.027 €**	**11.596.034**	**22%**	**6%**	**1,0**	**3,5**	**2,0**	**100%**	**C. Chance of good residual value without maintenance backlog**	
CG 500 Residual value	2,92%	203.154.417 €	168.674.979 €	194.560.972 €	34.479.438	8.593.445	20%	4%	1	3.5	2	100%	(1) Relationship between maintenance budget / useful life	
CG 500 Useful life		105	61	89	44	16	72%	18%					(cf. PPP School Study2019, margin no. 137) / (2) PPP/BKI: more favourable	
CG 500 Residual value risk (maintenance)*		14.937.825 €	-19.541.613 €	6.344.380 €	34.479.438	8.593.445	-176%	135%					construction costs lead to hidden reserves (3) Indexing residual value	
CG 500 Additional VAT revenue federal/state/local authorities	1,42%	3.002.589 €												
Life cycle costs excluding risk costs (Nominal)		**96.948.945 €**	**124.590.615 €**	**164.897.222 €**	**-27.641.671 €**	**-67.948.277 €**	**-22%**	**-41%**						
(NPV)		110.305.905 €	128.452.216 €	153.932.893 €	18.146.311 €	43.626.988 €	-14%	-28%						
Life cycle costs including risk costs (Nominal)		**84.453.254 €**	**147.425.290 €**	**158.552.842 €**	**-62.972.036 €**	**-74.099.589 €**	**-43%**	**-47%**	**1,8**	**3,3**	**2,5**	**(A+B+C)/3**	**PPP incentive structures promote quality management: cost caps, bonus/malus, savings participation, operator liability**	**1,4**
(NPV)		105.888.671 €	137.606.170 €	151.389.561 €	-31.717.500 €	-45.500.891 €	-23%	-30%						
Life cycle costs incl. risk costs + additional VAT revenue (Nominal)		**81.450.665 €**	**147.425.290 €**	**158.552.842 €**	**-65.974.624 €**	**-77.102.177 €**	**-45%**	**-49%**						
(NPV)		103.957.094 €	137.606.170 €	151.389.561 €	-33.649.076 €	-47.432.467 €	-24%	-31%						

* Nominal values indexed over 25 years, base year 2009 / Net present value (NPV) - discount rate 4.93 %. [2] Basis KG 310: 1 = MAX, 2 = PRE. Supply costs
[3] Quality assessment: 1 = very good / 2 = good / 3 = satisfactory / 4 = sufficient / 5 = unsatisfactory / [4] Questionnaire Construction and operating performance

Economic efficiency study — PPP project no 4 (greenfield / education sector)	Index	PPP[1]	KGSt[1]	BKI[1]	PPP - KGSt	PPP - BKI	PPP/ KGSt	PPP/ BKI	Quality[2] PPP	KGSt	BKI	Value	Remarks	FB B+B[4]
DIN 276 Construction costs									**2,2**	**2,8**	**2,8**	**100%**	**A. Construction quality, cost and schedule efficiency**	
DIN 277 GFA in m²		10.703	10.703	10.703										
Construction time in months		16	26	26	-10	-10	-38%	-38%	1	3	3	15%	Schedule reliability: N N	1
DIN 276 Construction costs excl. CG 760		24.114.395 €	25.552.412 €	25.552.412 €	-1.438.017	-1.438.017	-5.6%	-6%	2.75	2.75	2.75	70%	Building standard slightly above average (CG 300+400 passive house)	2.5
Construction costs /sqm GFA in T€		2.253 €	2.387 €	2.387 €	-134	-134	-6%	-6%					PPP cost certainty: N N	
CG 760 Interim Financing		626.907 €	814.979 €	814.979 €	-188.072	-188.072	-23%	-23%	1	3	3	15.0%	Interest rate hedging promotes high cost and deadline security	
Interest rate		3.701%	3.101%	3.101%	0.60%	0.60%	19%	19%						
DIN 276 Construction costs incl. CG 760		24.741.302 €	26.367.391 €	26.367.391 €	-1.626.089	-1.626.089	-6,2%	-6%						
Construction costs /sqm GFA in T€		2.312 €	2.463 €	2.463 €	-152	-152	-6%	-6%						
DIN 18960 Usage Costs excluding risk costs		**70.392.318**	**58.812.660**	**78.174.312**	**11.579.659**	**-7.781.994**	**20%**	**-10%**	**2,3**	**3,3**	**2,6**	**100%**	**B. High-quality maintenance/energy/property management**	
Usage costs including risk costs		**70.392.318**	**60.028.127**	**78.174.312**	**10.364.191**	**-7.781.994**	**17%**	**-10%**						
CG 100 Capital costs		36.740.036 €	37.951.584 €	37.951.584 €	-1.211.547	-1.211.547	-3%	-3%	3	3	3	14.3%	PPP: Forfaiting with waiver of defence	1
CG 100 Interest		11.998.734 €	11.584.193 €	11.584.193 €	414.541	414.541	4%	4%						
CG 100 Interest rates		2.830%	2.590%	2.590%	0.24%	0.24%	9%	9%						
CG 200 Property management	1.42%	6.057.031 €	3.982.174 €	3.437.184 €	2.074.857	2.619.847	52%	76%	1	3	2	14.3%	High-quality property management additional costs PFI promote QM	1
CG 210 Personnel costs	1.42%	5.858.699 €											VAT (Personnel costs) / CAFM / Quality assurance/ Remuneration of personnel	
CG 210 Own costs of the city	1.42%	198.332 €												
CG 300 Operating costs		15.564.158 €	12.002.198 €	23.845.688 €	3.561.960 €	8.281.530 €	30%	-35%						
CG 310 Supply costs [2]		2.686.091 €	4.033.970 €	5.673.798 €	1.347.879 €	2.987.707 €	-33%	-53%	2	3	4	14.3%	Cost advantage speaks in favour of lower energy consumption	1
CG 311 Water	1.49%	108.715 €	165.436 €	232.687 €	-56.721	-123.972	-34%	-53%					as yet no benchmark comparison of consumption quantities	
CG 312 Heating	1.12%	1.380.451 €	2.114.997 €	2.974.753 €	-734.546 €	1.594.302 €	-35%	-54%						
CG 313 Electricity	3.57%	1.196.925 €	1.753.537 €	2.466.358 €	-556.612	-1.269.433	-32%	-51%						
CG 320 Waste disposal	1.42%	599.992 €	599.992 €	599.992 €	0	0	0%	0%	3	3	3	14.3%	No qualitative study to date	1
CG 330 Cleaning	2.49%	10.011.892 €	6.324.633 €	5.745.230 €	3.687.259	4.266.662	58%	74%	3	3	3	14.3%	No qualitative study to date	2
CG 351 Energy management	2.20%				0	0								
CG 352 Inspection and maintenance	2.20%		469.655 €	10.681.322 €	-469.655	-10.681.322			1	5	1		PPP incentive structure: service levels and operator liability force	
CG 370 Taxes and insurance	1.56%		218.425 €	888.075 €	-218.425	-888.075			3	3	3		high-quality maintenance & inspection	
CG 370 Due Diligence	1.56%													
CG 391 Cafeteria operation	1.42%	218.434 €	218.434 €	218.434 €	0	0	0%	0%	3	3	3			
CG 393 Other operating costs	1.42%	2.047.750 €	137.089 €	38.838 €	1.910.661	2.008.912	1394%	5173%	2	3	3	14.3%	Additional costs promote quality management	
CG 400 Repair costs	2.92%	12.031.093 €	4.876.705 €	12.939.857 €	7.154.388 €	908.764 €	147%	-7%					Good maintenance management. Service levels, user requests for	
CG 460 Risiko Bauschäden (Instandhaltung)*	2.92%		1.215.468 €		-1.215.468	1.215.468							compliance with response/repair times. Cost ceilings, component-specific	
CG 200/350/400 Maintenance	2.92%	16.013.267 €	5.346.360 €	23.621.178 €	10.666.907	-7.607.912	200%	-32%	2	4	1	14.3%	maintenance calculation eduction in charges for poor performance.	1.6
CG 200/350/400 In % of restoration costs p.a. / CGST-Target 1.21%		1,66%	0,58%	2.39%									reserve account; CPP: damage risks due to low maintenance budgets	
Revenue / residual value excluding risk costs		**37.983.361 €**	**37.983.361 €**	**37.983.361 €**	**0 €**	**0**	**0%**	**0%**						
Revenue / residual value including risk costs		**40.738.199 €**	**27.624.263 €**	**44.198.821 €**	**13.113.936 €**	**-3.460.622**	**47%**	**-8%**	**2,0**	**4,0**	**1,0**	**100%**	**C. Chance of good residual value without maintenance backlog**	
Revenue / residual value including risk costs and VAT		**41.673.621 €**	**27.624.263 €**	**44.198.821 €**	**14.049.358 €**	**-2.525.199**	**51%**	**-6%**						
CG 500 Residual value including maintenance risk	2.92%	40.738.199 €	27.624.263 €	44.198.821 €	13.113.936	-3.460.622	47%	-8%	2	4	1	100%	(1) Relationship between maintenance budget / useful life	1
Useful life		91	56	110	36	-19	65%	-17%					(cf. PPP School Study2019, margin no. 137) / (2) PPP/BKI: more favourable	
Residual value risk (maintenance)*		2.754.837 €	-10.359.099 €	6.215.459 €	13.113.936	-3.460.622	-127%	-56%					construction costs lead to hidden reserves (3) Indexing residual value	
CG 500 Additional VAT revenue federal/state/local authorities		935.423 €												
Life cycle costs excluding risk costs Nominal		**32.408.957 €**	**20.829.298 €**	**40.190.951 €**	**11.579.659 €**	**-7.781.994 €**	**56%**	**-19%**						
NPV		30.663.139 €	22.885.390 €	35.950.965 €	7.777.750 €	-5.287.825 €	34%	-15%						
Life cycle costs including risk costs* Nominal		**29.654.119 €**	**32.403.864 €**	**33.975.492 €**	**-2.749.745 €**	**-4.321.372 €**	**-8%**	**-13%**	**2,2**	**3,4**	**2,1**	**(A+B+C) /3**	**PPP incentive structures promote quality management: cost caps, bonus/malus, savings participation, operator liability**	**1,3**
NPV		29.350.788 €	26.827.170 €	30.320.746 €	2.523.618 €	-969.958 €	9%	-3%						
Life cacle costs incl. risk costs + additional VAT revenue Nominal		**28.718.697 €**	**32.403.864 €**	**33.975.492 €**	**-3.685.167 €**	**-5.256.795 €**	**-11%**	**-15%**						
NPV		28.653.155 €	26.827.170 €	30.320.746 €	1.825.985 €	-1.667.591 €	7%	-5%						

1) Nominal values indexed over 30 years, base year 2014. Net present value (NPV) - discount rate 2.59 %. 2) CG 310 1 = MAX, 2 = PRE. Supply costs
3) Quality assessment 1 = very good / 2 = good / 3 = satisfactory / 4 = sufficient / 5 = unsatisfactory / 4) Questionnaire Construction and operating performance

	Economic efficiency study / PPP project no 5 (greenfield / education sector)	Index	PPP[1]	KGSt[1]	BKI[1]	PPP - KGSt	PPP - BKI	PPP/ KGSt	PPP/ BKI	Quality[3] PPP	KGSt	BKI	Value	Quality[3]	QCO[4]
DIN 276	**Construction costs**									**2,6**	**3,0**	**3,0**	**100%**	**A. Construction quality, cost and schedule efficiency**	
DIN 277	GFA in m2		37.117	37.117	37.117										
	Construction time in months		17	26	26	-9	-9	-35%	-35%	1	3	3	10%	Actual time 21 months due to expansion underground car park from 24 to 170 parking spaces	1
DIN 276	Construction costs excl. CG 760		38.269.007 €	52.383.310 €	52.383.310 €	-14.114.302	-14.114.302	-27%	-27%	3	3	3	70%	Medium building standard	2
	Construction costs /sqm GFA in T€		1.031 €	1.411 €	1.411 €	-380	-380	-27%	-27%	3	3	3	10%	PPP Cost certainty +7% (additional user request)	
CG 760	Interim Financing		748.514 €	1.058.694 €	1.058.694 €	-310.180	-310.180	-29%	-29%						
	Interest rate		3.458%	2.698%	2.698%	0.76%	0.76%	28%	28%	1	3	3	10%	Interest rate hedging promotes high cost and deadline security	
DIN 276	Construction costs incl. CG 760		39.017.521 €	53.442.004 €	53.442.004 €	-14.424.483	-14.424.483	-27%	-27%						
	Construction costs /sqm GFA in T€		1.051 €	1.440 €	1.440 €	-389	-389	-27%	-27%						
DIN 18960	**Usage Costs excluding risk costs**		103.837.807	114.345.013	129.454.231	-10.507.206	-25.616.424	-9%	-20%	**2,3**	**3,3**	**2,5**	**100%**	**B. High-quality maintenance/energy/property management**	
	Usage costs including risk costs*		103.837.807	115.667.748	129.454.231	-11.829.942	-25.616.424	-10%	-20%						1
CG 100	Capital costs		63.630.500 €	82.512.265 €	82.512.265 €	-18.881.765	-18.881.765	-23%	-23%	2	3	3	10%	Project finance (PFI): higher interest rates serve quality management (QM)	
CG 100	Interest		29.948.197 €	36.377.869 €	36.377.869 €	-6.429.672	-6.429.672	-18%	-18%						
CG 100	Interest rates		5.88%	5.28%	5.28%	0.60%	0.60%	11%	11%						
CG 200	Property management	1.42%	8.065.172 €	7.626.632 €	6.268.645	438.540	1.796.526	6%	29%	1	3	3	10%	High-quality property management, additional costs PFI promote QM	1
CG 210	Personnel costs	1.42%	7.684.602 €											VAT (Personnel costs) / CAFM / Quality assurance/ Remuneration of personnel	
CG 210	Own costs of the city	1.42%	380.570 €												
CG 300	Operating costs		25.895.103 €	15.976.351 €	23.900.167 €	9.918.752 €	1.994.937 €	62%	8%						
CG 310	Supply costs 2)		9.271.494 €	4.853.161 €	5.827.945 €	4.418.333 €	3.443.548 €	91%	59%	3	3	3	10%	No qualitative study to date	1
CG 311	Water	1.49%	867.761 €	144.257 €	181.409 €	723.505 €	686.353 €	502%	378%						
CG 312	Heating	1.12%	3.135.771 €	3.003.034 €	3.929.802 €	132.736 € -	794.032 €	4%	-20%						
CG 313	Electricity	3.57%	5.267.961 €	1.705.870 €	1.716.734 €	3.562.092 €	3.551.227 €	209%	207%						
CG 320	Waste disposal	1.42%	1.238.823 €	1.238.823 €	1.238.823 €	- €	- €	0%	0%	3	3	3	10%	No qualitative study to date	1
CG 330	Cleaning	2.49%	8.341.285 €	8.130.213 €	7.869.541 €	211.072 €	471.743 €	3%	6%	3	3	3	10%	No qualitative study to date	2
CG 350	Inspection and maintenance	2.20%	4.832.428 €	564.189 €	6.664.195 €	4.268.240 € -	1.831.767 €	757%	-27%	1	5	1	10%	PPP incentive structure High quality due to service levels and operator liability	
CG 370	Taxes and insurance	1.56%	1.751.393 €	406.414 €	1.578.304 €	1.344.980 €	173.089 €	331%	11%	2	3	2	10%	Additional costs PFI promote quality management	
CG 390	Other operating costs	1.42%	459.681 €	783.552 €	721.358 € -	323.871 € -	261.677 €	-41%	-36%	3	3	3	10%		
CG 391	Cafeteria operation	1.42%	459.681 €	459.681 €	459.681 €	- €	- €	0%	0%	3	3	3	10%		3
CG 393	Letter of comfort*	1.42%													
CG 393	Other operating costs	1.42%		323.871 €	261.677 € -	323.871 € -	261.677 €								
CG 400	Repair costs	2.92%	6.247.032 €	8.229.765 €	16.773.154 €	-1.982.733 €	-10.526.122 €	-24%	-63%					Good maintenance management. Service levels, user requests for compliance with response/repair times. Cost ceilings, component-specific maintenance calculation eduction in charges for poor performance reserve account; CPP: damage risks due to low maintenance budgets	
CG 460	Risk of structural damage (maintenance)*	2.92%		1.322.735 €	-	1.322.735 €	1.322.735 €								
CG 200/350/400	Maintenance		13.211.938 €	10.096.064 €	26.459.507 €	3.115.874	-13.247.568	31%	-50%	2	4	1	10%		1
CG 200/350/400	in % of restoration costs p.a. / CGST-Target 1.07%		1.40%	0.79%	1.97%										
Revenue	**Risiko Bauschäden (Instandhaltung)***		63.933.922 €	61.961.533 €	61.961.533 €	1.972.390 €	1.972.390 €	3%	0%	**2**	**4**	**1**		**C. Chance of good residual value without maintenance backlog**	
Revenue	**/ residual value including risk costs**		65.791.715 €	57.481.439 €	66.579.475 €	8.310.276 €	-787.760 €	14%	-1%						1
Revenue	**/ residual value including risk costs and VAT**		67.411.812 €	57.481.439 €	66.579.475 €	9.930.373 €	832.337 €	17%	1%						
CG 500	Residual value	2.92%	65.791.715 €	57.481.439 €	66.579.475	8.310.276	-787.760	14%	-1%					(1) Relationship between maintenance budget / useful life	
CG 500	Useful life (standard useful life 80 years)		87	67	100	20 -	13 €	30%	-13%					(cf. PPP School Study2019, margin no. 137) / (2) PPP/BKI more favourable	
CG 500	Residual value risk (maintenance)*		1.857.793 €	-4.480.094 €	4.617.943 €	6.337.887	-2.760.150	-141%	-60%					construction costs lead to hidden reserves (3) indexing residual value	
CG 500	Additional VAT revenue federal/state/local authorities	1.42%	1.620.097 €												
Life cycle costs excluding risk costs	Nominal		39.903.884 €	52.383.480 €	67.492.698 €	-12.479.596 €	-27.588.814 €	-24%	-41%						
	NPV		41.402.784 €	52.235.536 €	58.501.016 €	-10.832.852 €	-17.098.233 €	-21%	-29%						
Life cycle costs including risk costs*	Nominal		38.046.091 €	58.186.309 €	62.874.755 €	-20.140.218 €	-24.828.664 €	-35%	-39%	**2,3**	**3,4**	**2,2**	**(A+B+C)/3**	**PPP incentive structures promote quality management: cost caps, bonus/malus, savings participation, operator liability**	**1,4**
	NPV		40.703.866 €	54.418.708 €	56.763.708 €	-13.714.842 €	-16.059.842 €	-25%	-28%						
Life cycle costs incl. risk costs + additional VAT revenue	Nominal		36.425.995 €	58.186.309 €	62.874.755 €	-21.760.315 €	-26.448.761 €	-37%	-42%						
	NPV		39.690.249 €	54.418.708 €	56.763.708 €	-14.728.459 €	-17.973.459 €	-27%	-30%						

[1] Nominal values indexed over 20 years, base year 2009. Net present value (NPV) - discount rate 5.28 %   [2] CG 310 1 = MAX. 2 = PRE. Supply costs

[3] Quality assessment: 1 = very good / 2 = good / 3 = satisfactory / 4 = sufficient / 5 = unsatisfactory / [4] Questionnaire Construction and operating performance

Economic efficiency study — PPP project no 6 (greenfield / administration sector)	Index	PPP[1]	KGSt[1]	BKI[1]	PPP - KGSt	PPP - BKI	PPP/ KGSt	PPP/ BKI	Quality[3] PPP	KGSt	BKI	Value	Remarks	QCO[4]
DIN 276 Construction costs									**2,1**	**2,7**	**2,7**	**100%**	**A. Construction quality, cost and schedule efficiency**	
DIN 277 GFA in m2		11.168	11.168	11.168										
Construction time in months		20	26	26	-6	-6	-23%	-23.1%	1	3	3	10%	Schedule reliability 100%	2
DIN 276 Construction costs excl. CG 760		16.965.609 €	21.868.984 €	21.868.984 €	-4.903.374	-4.903.374	-22%	-22%	2,5	2,5	2,5	70%	Building standard above average; 2008 European Architecture Prize	1
Construction costs /sqm GFA in T€		1.519 €	1.958 €	1.958 €	-439	-439	-22%	-22%						
CG 760 Interim Financing		702.557 €	538.632 €	538.632 €	163.925	163.925	30%	30%	1	3	3	10%	Cost certainty 100%	
Interest rate		4.076%	3,476%	3,476%	0.60%	0.60%	17%	17%	1	3	3	10%	Interest rate hedging promotes high cost and deadline security	
DIN 276 Construction costs incl. CG 760		17.668.166 €	22.407.616 €	22.407.616 €	-4.739.449	-4.739.449	-21%	-21%						
Construction costs /sqm GFA in T€		1.582 €	2.006 €	2.006 €	-424	-424	-21%	-21%						
DIN 18960 Usage Costs excluding risk costs		**51.460.817**	**49.825.082**	**53.683.032**	**1.635.735**	**-2.222.215**	**3%**	**-4%**	**2,4**	**3,3**	**2,7**	**100%**	**B. High-quality maintenance/energy/property management**	
Usage costs including risk costs*		**51.460.817**	**50.719.512**	**53.683.032**	**741.304**	**-2.222.215**	**1%**	**-4%**						
CG 100 Capital costs		28.976.363 €	36.146.708 €	36.146.708 €	-7.170.345	-7.170.345	-20%	-20%	3	3	3		PPP: Forfaiting with waiver of defence	1
CG 100 Interest		11.308.197 €	13.739.092 €	13.739.092 €	-2.430.895	-2.430.895	-18%	-18%				10.0%		
CG 100 Interest rates		4.338%	4,178%	4,178%	0.16%	0.16%	4%	4%						
CG 200 Property management	1.42%	2.607.230 €	2.377.703 €	2.513.994 €	229.527	93.236	10%	4%	1	3	1	10.0%	High-quality property management; additional costs PFI promote QM	1
CG 210 Personnel costs	1.42%	2.441.118 €											VAT (Personnel costs) / CAFM / Quality assurance/ Remuneration	
CG 210 Own costs of the city	1.42%	166.111 €											of personnel	
CG 300 Operating costs		13.947.949 €	8.102.640 €	11.692.583 €	5.845.310 €	2.255.366 €	72%	19%						
CG 310 Supply costs 2)		2.788.275 €	3.188.034 €	3.728.856 € -	399.758 € -	940.580 €	-13%	-25%	2	3	4	10.0%	So far no comparison with benchmark (VDI)	1
CG 311 Water	1.49%	72.684 €	91.034 €	106.477 €	-18.349	-33.792	-20%	-32%					Consumption 2006-2020: 18% < guaranteed maximum quantity	
CG 312 Heating	1.12%	685.733 €	1.388.070 €	1.623.544 € -	702.337 € -	937.811 €	-51%	-58%					Consumption 2006-2020: 42% < guaranteed maximum quantity	
CG 313 Electricity	3.57%	2.029.868 €	1.708.930 €	1.998.835 €	320.928	31.023	19%	2%					Consumption 2006-2020: 19% < guaranteed maximum quantity	
CG 320 Waste disposal	1.42%	325.681 €	325.681 €	325.681 €	0	0	0%	0%	3	3	3	10.0%	No qualitative study to date	1
CG 330 Cleaning	2.49%	3.844.184 €	3.169.363 €	2.189.796 €	674.820	1.654.387	21%	76%	3	3	3	10.0%	No qualitative study to date	2
CG 352 Inspection and maintenance	2.20%	3.290.640 €	475.977 €	5.060.540 €	2.814.662	-1.769.901	591%	-35%	2	5	1	10.0%	PPP incentive structure: service levels and operator liability force	
CG 370 Taxes and insurance	1.56%	786.655 €	119.558 €	205.759 €	667.098	580.896	558%	282%	3	3	1	10.0%	No qualitative study to date	
CG 370 Due Diligence	1.56%													
CG 391 Cafeteria operation	1.42%	164.632 €	164.632 €	164.632 €	0	0	0%	0%	3	3	3	10.0%	No qualitative study to date	3
CG 393 Other operating costs	1.42%	2.747.882 €	659.394 €	17.318 €	2.088.488	2.730.564	317%	15767%	2	3	3	10.0%	Quality assurance by project company	
CG 400 Repair costs	2.92%	5.929.274 €	3.198.031 €	3.329.747 €	2.731.243 €	2.599.528 €	85%	78%					Good maintenance management. Service levels, user requests for	
CG 460 Risk of structural damage (maintenance)*	2.92%		894.431 €		-894.431	894.431							compliance with response/repair times, Cost ceilings, component-specific	
CG 200/350/400 Maintenance	2.92%	11.597.617 €	3.674.008 €	8.390.287 €	7.923.609	3.207.330	216%	38%	2	4	3	10.0%	maintenance calculation eduction in charges for poor performance,	1
CG 200/350/400 in % of restoration costs p.a. / CGST-Target	1.15%	1.70%	0,57%	1.19%									reserve account; CPP: damage risks due to low maintenance budgets	
Revenue / residual value excluding risk costs		**30.746.052 €**	**30.746.052 €**	**30.746.052 €**	**0 €**	**0**	**0%**	**0%**						
Revenue / residual value including risk costs		**32.699.614 €**	**24.756.562 €**	**30.918.589 €**	**7.943.052 €**	**1.781.025**	**32%**	**6%**	**1,5**	**4,0**	**3,0**	**100%**	**C. Chance of good residual value without maintenance backlog**	
Revenue / residual value including risk costs and VAT		**33.080.763 €**	**24.756.562 €**	**30.918.589 €**	**8.324.202 €**	**2.162.174**	**34%**	**7%**						
CG 500 Residual value	2.92%	32.699.614 €	24.756.562 €	30.918.589 €	7.943.052	1.781.025	32%	6%	1,5	4	3		(1) Relationship between maintenance budget / useful life	1
CG 500 Useful life (standard useful life 80 years)		93	56	81	37	12	66%	15%					(cf. PPP School Study2019, margin no. 137) / (2) PPP/BKI: more favourable	
CG 500 Residual value risk (maintenance)*		1.953.561 €	-5.989.491 €	172.537 €	7.943.052	1.781.025	-133%	1032%					construction costs lead to hidden reserves (3) Indexing residual value	
CG 500 Additional VAT revenue federal/state/local authorities		381.149 €												
Life cycle costs excluding risk costs (Nominal)		**20.714.764 €**	**19.079.029 €**	**22.936.979 €**	**1.635.735 €**	**-2.222.215 €**	**9%**	**-10%**						
(NPV)		21.330.070 €	19.916.255 €	22.267.033 €	1.413.815 €	-936.963 €	7%	-4%						
Life cycle costs including risk costs* (Nominal)		**18.761.203 €**	**25.962.951 €**	**22.764.443 €**	**-7.201.748 €**	**-4.003.240 €**	**-28%**	**-18%**	**2,0**	**3,3**	**2,8**	**(A+B+C)/3**	**PPP incentive structures promote quality management: cost caps, bonus/malus, savings participation, operator liability**	**1,3**
(NPV)		20.598.590 €	22.493.831 €	22.202.430 €	-1.895.241 €	-1.603.840 €	-8%	-7%						
Life cycle costs incl. risk costs + additional VAT revenue (Nominal)		**18.380.053 €**	**25.962.951 €**	**22.764.443 €**	**-7.582.897 €**	**-4.384.389 €**	**-29%**	**-19%**						
(NPV)		20.598.352 €	22.493.831 €	22.202.430 €	-1.895.479 €	-1.604.077 €	-8%	-7%						

[1] Nominal values indexed over 25 years, base year 2006. Net present value (NPV) - discount rate 3.47 %. 2) CG 310 1 = MAX, 2 = PRE. Supply costs
[3] Quality assessment: 1 = very good / 2 = good / 3 = satisfactory / 4 = sufficient / 5 = unsatisfactory / 4) Questionnaire Construction and operating performance

Economic efficiency study — PPP project no 7 (greenfield / education sector)		Index	PPP[1]	KGSt[1]	BKI[1]	PPP - KGSt	PPP - BKI	PPP/ KGSt	PPP/ BKI	Quality[3] PPP	KGSt	BKI	Value	Remarks	QCO[4]
DIN 276	**Construction costs**									**2,4**	**3,0**	**3,0**	**100%**	**A. Construction quality, cost and schedule efficiency**	
DIN 277	GFA in m² high school		11.600	11.600	11.600										
DIN 277	GFA in m² (High school and sports hall)		13.743	13.743	13.743										
	Construction time in months		17	26	26	-9	-9	-35%	-35%	1	3	3	10%	Date certainty 100% (2 months before target date)	2
DIN 276	Construction costs excl. CG 760		21.684.815 €	23.768.552 €	23.768.552 €	-2.083.736	-2.083.736	-8,8%	-9%	3	3	3	70%	Medium building standard	2
	Construction costs /sqm GFA in T€		1.869 €	2.049 €	2.049 €	-180	-180	-9%	-9%	1	3	3	10%	PPP Cost certainty 100% (14 T€ < Target)	
CG 760	Interim Financing		182.448 €	263.575 €	263.575 €	-81.127	-81.127	-31%	-31%						
	Interest rate		2.596%	1.996%	1.996%	0.60%	0.60%	30%	30%	1	3	3	10%	Interest rate hedging promotes high cost and deadline security	
DIN 276	Construction costs incl. CG 760		21.867.264 €	24.032.127 €	24.032.127 €	-2.164.863	-2.164.863	-9,0%	-9%						
	Construction costs /sqm GFA in T€		1.885 €	2.072 €	2.072 €	-187	-187	-9%	-9%						
DIN 18960	**Usage Costs excluding risk costs**		42.877.431	41.372.374	47.269.243	1.505.056	-4.391.813	4%	-9%	**2,4**	**3,3**	**2,8**	**100%**	**B. High-quality maintenance/energy/property management**	
	Usage costs including risk costs*		42.877.431	41.819.349	47.269.243	1.058.082	-4.391.813	3%	-9%						
CG 100	Capital costs		26.057.051 €	28.636.704 €	28.636.704 €	-2.579.653	-2.579.653	-9%	-9%	3	3	3	10%	PPP final financing via local authority	2
CG 100	Interest (incl. IZZB)		4.189.788 €	4.604.577 €	4.604.577 €	-414.790	-414.790	-9%	-9%						
CG 100	Interest rates		2.00%	2.00%	2.00%	0.00%	0.00%	0%	0%						
CG 200	Property management	1.42%	5.044.471 €	2.402.224 €	2.058.932 €	2.642.247	2.985.539	110%	145%	1	3	3	10%	High-quality property management, additional costs PFI promote QM	2
CG 210	Personnel costs	1.42%	4.808.528 €	2.402.224 €	2.058.932 €									VAT (Personnel costs) / CAFM / Quality assurance/ Remuneration of	
CG 210	Own costs of the city	1.42%	235.943 €											personnel	
CG 300	Operating costs		7.778.610 €	7.641.139 €	10.476.135 €	137.470	-2.697.526	2%	-26%						
CG 310	Supply costs 2)		2.027.511 €	2.318.563 €	3.193.765 €	-291.052	-1.166.254	-13%	-37%	2	3	4	10%	So far no comparison with benchmark (VDI)	2
CG 311	Water	1.49%	40.280 €	93.837 €	134.313 €	-53.558	-94.034	-57%	-70%					Consumption 2016-2020: 65% < guaranteed maximum quantity	
CG 312	Heating	1.12%	849.388 €	1.291.947 €	1.526.517 €	-442.559	-677.129	-34%	-44%					Consumption 2016-2020: 41% < guaranteed maximum quantity	
CG 313	Electricity	3.57%	1.137.843 €	932.778 €	1.532.935 €	205.065	-395.091	22%	-26%					Consumption 2016-2020: 26% < guaranteed maximum quantity	
CG 320	Waste disposal	1.42%	361.942 €	361.942 €	361.942 €	0	0	0%	0%	3	3	3	10%	No qualitative study to date	2
CG 330	Cleaning	2.49%	2.899.907 €	4.246.100 €	3.904.902 €	-1.346.193	-1.004.995	-32%	-26%	3	3	3	10%	No qualitative study to date	5
CG 350	Inspection and maintenance	2.20%	2.211.158 €	270.826 €	2.186.996 €	1.940.332	24.162	716%	1%	1	5	1	10%	PPP incentive structure. High quality due to service levels and operator liability	
CG 370	Taxes and insurance	1.56%	0 €	130.671 €	527.174 €	-130.671	-527.174			3	3	3	10%	No qualitative study to date	
CG 391	Cafeteria operation	1.42%	278.092 €	278.092 €	278.092 €	0	0	0%	0%	3	3	3	10%	No qualitative study to date	
CG 393	Other operating costs	1.42%	0 €	34.946 €	23.265 €	-34.946	-23.265			3	3	3	10%	No qualitative study to date	
CG 400	Repair costs	2.92%	3.997.299 €	3.139.281 €	6.097.472 €	858.018	-2.100.173	27%	-34%					Good maintenance management. Service levels, user requests for	
CG 460	Risk of structural damage (maintenance)*	2.92%		446.974 €		-446.974	446.974							compliance with response/repair times. Cost ceilings, component-specific	
CG 200/350/400	Maintenance		7.407.964 €	3.589.015 €	9.041.954 €	3.818.948	-1.633.990	106%	-18%	2	4	2	10%	maintenance calculation eduction in charges for poor performance	2
CG 200/350/400	in % of restoration costs p.a. / CGST-Target	1.15%	1.62%	0.67%	1.68%									reserve account; CPP: damage risks due to low maintenance budgets	
	Revenue / residual value excluding risk costs		30.386.426 €	30.386.426 €	30.386.426 €	0 €	0	0%	0%						
	Revenue / residual value including risk costs		31.452.806 €	27.445.805 €	31.619.590 €	4.007.001 €	-166.785	15%	-1%	**2,0**	**4,0**	**2,0**	**100%**	**C. Chance of good residual value without maintenance backlog**	
	Revenue / residual value including risk costs and VAT		32.242.198 €	27.445.805 €	31.619.590 €	4.796.393 €	622.607	17%	2%						
CG 500	Residual value	2.92%	31.452.806 €	27.445.805 €	31.619.590 €	4.007.001	-166.785	15%	-1%	2	4	2	100%	(1) Relationship between maintenance budget / useful life	2
	Useful life (standard useful life 80 years)		91	60	93	31	-2	52%	-2%					(cf. PPP School Study2019, margin no 137) / (2) PPP/BKI: more favourable	
	Residual value risk (maintenance)*		1.066.379 €	-2.940.622 €	1.233.164 €	4.007.001	-166.785	-136%	-16%					construction costs lead to hidden reserves (3) Indexing residual value	
CG 500	Additional VAT revenue federal/state/local authorities	1.42%	789.392 €												
	Life cycle costs excluding risk costs (Nominal)		12.491.004 €	10.985.948 €	16.882.817 €	1.505.056 €	-4.391.813 €	14%	-26%						
	(NPV)		14.561 €	13.320 €	18.260 €	1.241 €	-3.698 €	9%	-20%						
	Life cycle costs including risk costs* (Nominal)		11.424.625 €	14.373.544 €	15.649.653 €	-2.948.920 €	-4.225.028 €	-21%	-27%	**2,3**	**3,4**	**2,6**	**(A+B+C)/3**	**PPP incentive structures promote quality management: cost caps, bonus/malus, savings participation, operator liability**	**2,3**
	(NPV)		13.800 €	15.740 €	17.379 €	-1.940 €	-3.579 €	-12%	-21%						
	Life cycle costs incl. risk costs + additional VAT revenue (Nominal)		10.635.233 €	14.373.544 €	15.649.653 €	-3.738.311 €	-5.014.420 €	-26%	-32%						
	(NPV)		13.134 €	15.740 €	17.379 €	-2.605 €	-4.245 €	-17%	-24%						

[1] Nominal values indexed over 18 years, base year 2015 (begin of usage), Net present value (NPV) - discount rate 2.0 %. [2] CG 310: CG 310 1 = MAX, 2 = PRE. Supply costs
[3] Quality assessment: 1 = very good / 2 = good / 3 = satisfactory / 4 = sufficient / 5 = unsatisfactory / [4] Questionnaire Construction and operating performance

Economic efficiency study
PPP project no 8 (greenfield / education sector)

		Index	PPP[1]	KGSt[1]	BKI[1]	PPP - KGSt	PPP - BKI	PPP/KGSt	PPP/BKI	Quality[3] PPP	KGSt	BKI	Value	Remarks	QCO[4]
DIN 276	**Construction costs**									**2,4**	**3,0**	**3,0**	**100%**	**A. Construction quality, cost and schedule efficiency**	
DIN 277	GFA in m2		15.956	15.956	15.956										
	Construction time in months		15	26	26	-11	-11	-41%	-41%	1	3	3	10%	Schedule reliability 100%	2
DIN 276	Construction costs excl. CG 760		22.952.542 €	30.042.447 €	30.042.447 €	-7 089 905	-7 089 905	-24%	-24%	3	3	3	70%	Medium building standard	2
	Construction costs /sqm GFA in T€		1.438 €	1.883 €	1.883 €	-444	-444	-24%	-24%	1	3	3	10%	PPP: High cost certainty (+1%)	
CG 760	Interim Financing		391.957 €	488.146 €	488.146 €	-96.189	-96.189	-20%	-20%						
	Interest rate		2.964%	2.364%	2.364%	0.60%	0.60%	25%	25%	1	3	3	10%	Interest rate hedging promotes high cost and deadline security	
DIN 276	Construction costs incl. CG 760		23.344.499 €	30.530.594 €	30.530.594 €	-7 186 095	-7 186 095	-24%	-24%						
	Construction costs /sqm GFA in T€		1.463 €	1.913 €	1.913 €	-450	-450	-24%	-24%						
DIN 18960	**Usage Costs excluding risk costs**		**60.258.145**	**67.313.566**	**84.099.199**	**-7.055.420**	**-23.841.053**	**-10%**	**-28%**	**2,6**	**3,3**	**2,6**	**100%**	**B. High-quality maintenance/energy/property management**	
	Usage costs including risk costs*		**60.258.145**	**68.532.236**	**84.099.199**	**-8.274.091**	**-23.841.053**	**-12%**	**-28%**						
CG 100	Capital costs		35.501.405 €	45.691.062 €	45.691.062 €	-10.189.657	-10.189.657	-22%	-22%	3	3	3	10.0%	PPP: Forfaiting with waiver of defence	2
CG 100	Interest		12 156 905 €	15 160 468 €	15 160 468 €	-3 003 563	-3 003 563	-20%	-20%						
CG 100	Interest rates		3.62%	3.47%	3.47%	0.15%	0.15%	4%	4%						
CG 200	Property management	1.42%	2.644.853 €	4.111.859 €	3.923.124 €	-1.467.006	-1.278.271	36%	33%	3	3	3	10.0%	VAT (Personnel costs) / CAFM / Quality assurance/ Remuneration of	3
CG 210	Personnel costs	1.42%	652 909 €											personnel	
CG 210	Own costs of the city	1.42%	1 991 944 €											Municipal caretaker for school-related services	
CG 300	Operating costs		14.762.538 €	12.302.454 €	22.184.329 €	2.460.085 €	7.421.791 €	20%	-33%						
CG 310	Supply costs 2) [2]	0.00%	2 975 279 €	4 767 551 €	4 798 709 €	1 792 272 €	1 823 430 €	-38%	-38%	2	3	3	10.0%	Cost advantage speaks in favour of lower energy consumption	3
CG 311	Water	1.49%	138 142 €	687 387 €	706 599 €	-549 244	-568 456	-80%	-80%					as yet no benchmark comparison of consumption quantities	
CG 312	Heating	1.12%	1 585 626 €	2 293 861 €	2 390 191 €	708 235 €	804 565 €	-31%	-34%					PPP company bears the risk of excess/shortfall consumption	
CG 313	Electricity	3.57%	1.251 511 €	1.786 304 €	1.701.920 €	-534.793	-450 408	-30%	-26%						
CG 320	Waste disposal	1.42%	761 985 €	761 985 €	761 985 €	0	0	0%	0%	3	3	3	10.0%	Not part of contract	
CG 330	Cleaning	2.49%	6.755.214 €	5 695.202 €	6 059.938 €	1.060 012	695.276	19%	11%	3	3	3	10.0%	No qualitative study to date	4
CG 351	Energy management	2.20%	247 409 €												
CG 352	Inspection and maintenance	2.20%	3.672 254 €	486.390 €	9.017.795 €	3.185 864	-5 345 541	655%	-59%	1	5	1	10.0%	PPP incentive structure. High quality due to service levels and operator liability	
CG 370	Taxes and insurance	1.56%	0 €	288 239 €	988 929 €	-288 239	-988 929			3	3	3	10.0%	No qualitative study to date	
CG 370	Due Diligence	1.56%	0 €												
CG 391	Cafeteria operation	1.42%	261 392 €	261 392 €	261 392 €	0	0	0%	0%	3	3	3	10.0%	No qualitative study to date	
CG 393	Other operating costs	1.42%	89 005 €	41.694 €	295.581 €	47.312	-206.575	113%	-70%	3	3	3	10.0%	No qualitative study to date	
CG 400	Repair costs	2.92%	7.349.349 €	5.208.192 €	12.300.683 €	2.141.158 €	4.951.334 €	41%	-40%					Good maintenance management. Service levels, user requests for	
CG 460	Risk of structural damage (maintenance)*	2.92%		1.218.670 €		-1.218.670	1.218.670							compliance with response/repair times, Cost ceilings, component-specific	
CG 200/350/400	Maintenance	2.92%	15 133 462 €	5 694 582 €	21 318 478 €	9 438 880	-6 185 016	166%	-29%	2	4	1	10.0%	maintenance calculation eduction in charges for poor performance	
CG 200/350/400	in % of restoration costs p.a. / CGST-Target 1.28%		1.74%	0.63%	2.32%									reserve account; CPP: damage risks due to low maintenance budgets	
	Revenue / residual value excluding risk costs		**41.891.794 €**	**41.891.794 €**	**41.891.794 €**	**0 €**	**0**	**0%**	**0%**						
	Revenue / residual value including risk costs		**44.553.541 €**	**33.731.055 €**	**46.828.538 €**	**10.822.486 €**	**-2.274.997**	**32%**	**-5%**	**2,0**	**4,0**	**1,0**	**100%**	**C. Chance of good residual value without maintenance backlog**	
	Revenue / residual value including risk costs and VAT		**44.693.545 €**	**33.731.055 €**	**46.828.538 €**	**10.962.490 €**	**-2.134.993**	**32%**	**-5%**						
CG 500	Residual value	2.92%	44.553.541 €	33.731.055 €	46.828.538 €	10.822.486	-2.274.997	32%	-5%	2	4	1		(1) Relationship between maintenance budget / useful life	3
	Useful life (standard useful life 80 years)		93	56	108	37	-15	66%	-14%					(cf. PPP School Study2019, margin no. 137) / (2) PPP/BKI: more favourable	
	Residual value risk (maintenance)*		2 661 746 €	-8 160 739 €	4 936 743 €	10.822.486	-2 274 997	-133%	-46%					construction costs lead to hidden reserves (3) Indexing residual value	
CG 500	Additional VAT revenue federal/state/local authorities		140 004 €												
	Life cycle costs excluding risk costs	Nominal	**18.366.351 €**	**25.421.771 €**	**42.207.404 €**	**-7.055.420 €**	**-23.841.053 €**	**-28%**	**-56%**						
		NPV	21.586.951 €	27.030.434 €	38.038.381 €	-5.443.483 €	-16.451.430 €	-20%	-43%						
	Life cycle costs including risk costs*	Nominal	**15.704.604 €**	**34.801.181 €**	**37.270.661 €**	**-19.096.576 €**	**-21.566.056 €**	**-55%**	**-58%**	**2,3**	**3,4**	**2,2**	**(A+B+C)/3**	**PPP incentive structures promote quality management: cost caps, bonus/malus, savings participation, operator liability**	**2,6**
		NPV	20.412.264 €	31.169.772 €	35.859.688 €	-10.757.508 €	-15.447.424 €	-35%	-43%						
	Life cycle costs incl. risk costs + additional VAT revenue	Nominal	**15.564.600 €**	**34.801.181 €**	**37.270.661 €**	**-19.236.581 €**	**-21.706.060 €**	**-55%**	**-58%**						
		NPV	20.412.171 €	31.169.772 €	35.859.688 €	-10.757.602 €	-15.447.517 €	-35%	-43%						

[1] Nominal values indexed over 25 years, base year 2012 / Net present value (NPV) - discount rate: 3.47 %; [2] CG 310: 1 = MAX, 2 = PRE: Supply costs
[3] Quality assessment: 1 = very good / 2 = good / 3 = satisfactory / 4 = sufficient / 5 = unsatisfactory / [4] Questionnaire: Construction and operating performance

Economic efficiency study — PPP project no 9 (greenfield / education sector)	Index	PPP[1]	KGSt[1]	BKI[1]	PPP - KGSt	PPP - BKI	PPP/ KGSt	PPP/ BKI	Q PPP	Q KGSt	Q BKI	Value	Remarks	QCO[4]
DIN 276 Construction costs									2,4	3,0	3,0	100%	A. Construction quality, cost and schedule efficiency	
DIN 277 GFA in m2		15.584	15.584	15.584										
Construction time in months		17	26	26	-9	-9	-35%	-35%	1	3	3	10%	Schedule reliability 100%	2
DIN 276 Construction costs excl. CG 760		23.771.674 €	30.004.080 €	30.004.080 €	-6.232.406	-6.232.406	-21%	-21%	3	3	3	70%	Medium building standard	2
Construction costs /sqm GFA in T€		1.525 €	1.925 €	1.925 €	-400	-400	-21%	-21%	1	3	3	10%	PPP: High cost certainty (+1.39% additional user requests)	
CG 760 Interim Financing		569.496 €	720.564 €	720.564 €	-151.068	-151.068	-21%	-21%						
Interest rate		3.701%	3.101%	3.101%	0.60%	0.60%	19%	19%	1	3	3	10%	Interest rate hedging promotes high cost and deadline security	
DIN 276 Construction costs incl. CG 760		24.341.170 €	30.724.644 €	30.724.644 €	-6.383.474	-6.383.474	-21%	-21%						
Construction costs /sqm GFA in T€		1.562 €	1.972 €	1.972 €	-410	-410	-21%	-21%						
DIN 18960 Usage Costs excluding risk costs		63.639.044	71.063.818	87.027.360	-7.424.774	-23.388.316	-10%	-27%	2,7	3,4	2,7	100%	B. High-quality maintenance/energy/property management	2
Usage costs including risk costs*		63.639.044	72.290.234	87.027.360	-8.651.190	-23.388.316	-12%	-27%						
CG 100 Capital costs		39.758.974 €	49.563.271 €	49.563.271 €	-9.804.297	-9.804.297	-20%	-20%	3	3	3	10.0%	PPP: Forfaiting with waiver of defence	
CG 100 Interest		15.417.804 €	18.838.627 €	18.838.627 €	-3.420.823	-3.420.823	-18%	-18%						
CG 100 Interest rates		4.338%	4.178%	4.178%	0.16%	0.16%	4%	4%						
CG 200 Property management	1.42%	3.383.133 €	4.058.478 €	3.858.602 €	-675.345	-475.469	-17%	-12%	3	3	3	10.0%	VAT (Personnel costs) / CAFM / Quality assurance/ Remuneration of	3
CG 210 Personnel costs	1.42%	1.437.679 €											personnel	
CG 210 Own costs of the city	1.42%	1.945.453 €											Municipal caretaker for school-related services	
CG 300 Operating costs		14.487.724 €	12.244.390 €	21.417.682 €	2.243.334 € -	6.929.958 €	18%	-32%						
CG 310 Supply costs 2)		2.816.126 €	4.784.279 €	4.744.478 € -	1.968.153 € -	1.928.352 €	-41%	-41%	2	4	4	10.0%	Cost advantage speaks in favour of lower energy consumption	3
CG 311 Water	1.49%	129.068 €	686.390 €	693.791 €	-557.322	-564.724	-81%	-81%					Local authority bears the risk of excess/shortfall consumption	
CG 312 Heating	1.12%	980.769 €	2.259.150 €	2.310.528 € -	1.278.381 € -	1.329.759 €	-57%	-58%					Heating consumption (2014-2020): Act. 35%< Target (MAX)	
CG 313 Electricity	3.57%	1.706.289 €	1.838.739 €	1.740.159 €	-132.450	-33.870	-7%	-2%					Electricity consumption (2014-2020): Actuel 3.7% < Target (MAX)	
CG 320 Waste disposal	1.42%	752.093 €	752.093 €	752.093 €	0	0	0%	0%	3	3	3	10.0%	Not part of contract	
CG 330 Cleaning	2.49%	7.099.336 €	5.665.975 €	5.991.834 €	1.433.361	1.107.502	25%	18%	3	3	3	10.0%	No qualitative study to date	4
CG 351 Energy management	2.20%	325.584 €												
CG 352 Inspection and maintenance	2.20%	2.700.364 €	467.346 €	8.401.895 €	2.233.018	-5.701.530	478%	-68%	1	5	1	10.0%	PPP incentive structure: High quality due to service levels and operator liability	
CG 370 Taxes and insurance	1.56%		278.253 €	973.375 €	-278.253	-973.375			3	3	3	10.0%	No qualitative study to date	
CG 370 Due Diligence	1.56%													
CG 391 Cafeteria operation	1.42%	255.292 €	255.292 €	255.292 €	0	0	0%	0%	3	3	3	10.0%	No qualitative study to date	
CG 393 Other operating costs	1.42%	538.929 €	41.152 €	298.715 €	497.776	240.214	1210%	80%	3	3	3	10.0%	PPP may include costs from other cost groups	
CG 400 Repair costs	2.92%	6.009.214 €	5.197.680 €	12.187.806 €	811.534 € -	6.178.592 €	16%	-51%					Good maintenance management: Service levels, user requests for compliance with response/repair times. Cost ceilings, component-specific	
CG 460 Risk of structural damage (maintenance)*	2.92%		1.226.416 €		-1.226.416	1.226.416							maintenance calculation eduction in charges for poor performance.	3
CG 200/350/400 Maintenance	2.92%	12.768.056 €	5.665.026 €	20.589.700 €	7.103.030	-7.821.644	125%	-38%	2.5	4	1	10.0%	reserve account. CPP: damage risks due to low maintenance budgets	
CG 200/350/400 in % of restoration costs p.a. / CGST-Target	1.24%	1.40%	0.63%	2.24%										
Revenue / residual value excluding risk costs		42.158.055 €	42.158.055 €	42.158.055 €	0 €	0	0%	0%	2,5	4,0	1,0	100%	C. Chance of good residual value without maintenance backlog	3
Revenue / residual value including risk costs		43.285.276 €	34.425.716 €	47.126.176 €	8.859.559 €	-3.840.900	26%	-8%						
Revenue / residual value including risk costs and VAT		43.561.878 €	34.425.716 €	47.126.176 €	9.136.162 €	-3.564.298	27%	-8%						
CG 500 Residual value	2.92%	43.285.276 €	34.425.716 €	47.126.176 €	8.859.559	-3.840.900	26%	-8%	2.5	4	1		(1) Relationship between maintenance budget / useful life	3
CG 500 Useful life (standard useful life 80 years)		85	57	108	28	-23	49%	-21%					(cf. PPP School Study2019, margin no. 137) / (2) PPP/BKI: more favourable	
CG 500 Residual value risk (maintenance)*		1.127.221 €	-7.732.339 €	4.968.121 €	8.859.559	-3.840.900	-115%	-77%					construction costs lead to hidden reserves (3) Indexing residual value	
CG 500 Additional VAT revenue federal/state/local authorities	1.42%	276.603 €												
Life cycle costs excluding risk costs (Nominal)	Nominal	21.480.989 €	28.905.763 €	44.869.305 €	-7.424.774 €	-23.388.316 €	-26%	-52%						
	NPV	24.038.946 €	28.993.751 €	38.679.574 €	4.954.805 €	-14.640.628 €	-17%	-38%						
Life cycle costs including risk costs*	Nominal	20.353.768 €	37.864.518 €	39.901.184 €	-17.510.750 €	-19.547.416 €	-46%	-49%	2,5	3,5	2,2	(A+B+C)/3	PPP incentive structures promote quality management: cost caps, bonus/malus, savings participation, operator liability	2,6
	NPV	23.616.876 €	32.348.216 €	36.819.339 €	-8.731.340 €	-13.202.463 €	-27%	-36%						
Life cycle costs incl. risk costs + additional VAT revenue	Nominal	20.077.165 €	37.864.518 €	39.901.184 €	-17.787.352 €	-19.824.018 €	-47%	-50%						
	NPV	23.616.784 €	32.348.216 €	36.819.339 €	-8.731.511 €	-13.202.835 €	-27%	-36%						

** Nominal values indexed over 25 years, base year 2013 / Net present value (NPV) - discount rate 4.575 %　2) CG 310 1 = MAX 2 = PRE Supply costs

3) Quality assessment　1 = very good / 2 = good / 3 = satisfactory / 4 = sufficient / 5 = unsatisfactory / 4) Questionnaire Construction and operating performance

Economic efficiency study — PPP project no 10 (greenfield / education sector)		Index	PPP[1]	KGSt[1]	BKI[1]	PPP - KGSt	PPP - BKI	PPP/ KGSt	PPP/ BKI	Quality[3] PPP	KGSt	BKI	Value	Remarks	QCO[4]
DIN 276	Construction costs									2,4	3,0	3,0	100%	A. Construction quality, cost and schedule efficiency	
DIN 277	GFA in m²		29.818	29.818	29.818										
	Construction time in months		24	30	30	-6	-6	-19%	-19%	1	3	3	10%	Schedule certainty building 99% (+2 weeks); demolition/outdoor facilities 100%	1
DIN 276	Construction costs excl. CG 760		53.287.843 €	57.992.038 €	57.992.038 €	-4.704.194	-4.704.194	-8%	-8.1%	3	3	3	70%	Medium building standard	2
	Construction costs /sqm GFA in T€		1.787 €	1.945 €	1.945 €	-158	-158	-8%	-8%	1	3	3	10%	PPP Cost certainty 100%	
CG 760	Interim Financing		2.157.650 €	2.085.331 €	2.085.331 €	72.319	72.319	3%	3%						
	Interest rate		3.568%	2.908%	2.908%	0.66%	0.66%	23%	23%	1	3	3	10%	Interest rate hedging promotes high cost and deadline certainty	
DIN 276	Construction costs incl. CG 760		55.445.494 €	60.077.369 €	60.077.369 €	-4.631.875	-4.631.875	-8%	-8%						
	Construction costs /sqm GFA in T€		1.859 €	2.015 €	2.015 €	-155	-155	-8%	-7.7%						
DIN 18960	Usage Costs excluding risk costs		156.548.374	150.495.827	174.860.711	6.052.547	-18.312.336	4%	-10%	2,3	3,3	2,8	100%	B. High-quality maintenance/energy/property management	
	Usage costs including risk costs*		156.548.374	152.674.808	174.860.711	3.873.566	-18.312.336	3%	-10%						
CG 100	Capital costs		103.242.814 €	106.108.794 €	106.108.794 €	-2.865.980	-2.865.980	-3%	-3%	2	3	3	10.0%	Project finance (PFI): higher interest rates serve quality management (QM)	k.A.
CG 100	Interest		47.797.320 €	46.031.425 €	46.031.425 €	1.765.895	1.765.895	4%	4%						
CG 100	Interest rates		5.67%	5.07%	5.07%	0.60%	0.60%	12%	12%						
CG 200	Property management	1.42%	6.858.888 €	8.329.782 €	7.139.405 €	-1.470.894	-280.517	-18%	-4%	2	3	2.5	10.0%	VAT (Personnel costs) / CAFM / Quality assurance / Remuneration	1
CG 210	Personnel costs	1.42%	6.427.663 €												
CG 210	Own costs of the city	1.42%	431.224 €												
CG 300	Operating costs		24.698.479 €	26.954.709 €	36.574.661 €	2.256.230 €	11.876.182 €	-8%	-32%						
CG 310	Supply costs 2)		6.029.044 €	7.955.117 €	11.110.334 €	1.926.073 €	5.081.290 €	-24%	-46%	2	3	4	10.0%	Savings compared to guaranteed maximum quantities are shared	1
CG 311	Water	1.49%	406.373 €	324.947 €	688.354 €	81.426	-281.981	25%	-41%					Consumption 2010-2019: 3% < guaranteed maximum quantity	
CG 312	Heating	1.12%	1.779.935 €	4.505.132 €	4.011.894 €	2.725.197 €	2.231.959 €	-60%	-56%					Consumption 2010-2019: 36% < guaranteed maximum quantity	
CG 313	Electricity	3.57%	3.842.736 €	3.125.037 €	6.410.086 €	717.699	-2.567.350	23%	-40%					Consumption 2010-2019: 26% < guaranteed maximum quantity	
CG 320	Waste disposal	1.42%	1.255.043 €	1.255.043 €	1.255.043 €	0	0	0%	0%	3	3	3	10.0%	No qualitative study to date	1
CG 330	Cleaning	2.49%	8.048.385 €	12.197.683 €	10.944.643 €	-4.149.298	-2.896.258	-34%	-26%	3	3	3	10.0%	No qualitative study to date	1
CG 350	Inspection and maintenance	2.20%	4.503.131 €	926.147 €	7.478.892 €	3.576.983	-2.975.761	386%	-40%	1	5	1	10.0%	PPP incentive structure: High quality due to service levels and operator liability	
CG 370	Taxes and insurance	1.56%	496.062 €	451.862 €	1.822.979 €	44.200	-1.326.918	10%	-73%						
CG 370	Due Diligence	1.56%	302.660 €							2	3	3	10.0%	Additional PFI costs promote quality management	
CG 391	Cafeteria operation	1.42%	3.882.098 €	3.882.098 €	3.882.098 €	0	0	0%	0%	3	3	3	10.0%	No qualitative study to date	3
CG 393	Other operating costs	1.42%	182.056 €	286.758	80.671	-104.702	101.385	-37%	56%	3	3	3	10.0%		
CG 400	Repair costs	2.92%	21.748.194 €	9.102.543 €	25.037.851 €	12.645.651 €	3.289.657 €	139%	-13%	2.25	4	2	10.0%	Good maintenance management: Service levels, user requests for	
CG 460	Risk of structural damage (maintenance)*	2.92%		2.178.981 €		-2.178.981	2.178.981							compliance with response/repair times. Cost ceilings component-specific	
CG 200/350/400	Maintenance	2.92%	34.581.106 €	11.570.818 €	34.940.728 €	23.010.288	-359.623	199%	-1%					maintenance calculation eduction in charges for poor performance.	1
CG 200/350/400	in % of restoration costs p.a. / CGST-Target	1.14%	1.71%	0.61%	1.84%									reserve account: CPP damage risks due to low maintenance budgets	
Revenue / residual value excluding risk costs			82.433.667 €	82.433.667 €	82.433.667 €	0 €	0	0%	0%						
Revenue / residual value including risk costs			87.321.038 €	68.220.966 €	88.678.642 €	19.100.073 €	-1.357.603	28%	-2%	2.25	4,0	2,0	100%	C. Chance of good residual value without maintenance backlog	
Revenue / residual value including risk costs and VAT			88.583.370 €	68.220.966 €	88.678.642 €	20.362.404 €	-95.272	30%	0%						
CG 500	Residual value	2.92%	87.321.038 €	68.220.966 €	88.678.642 €	19.100.073	-1.357.603	28%	-2%	2.25	4	2	100%	(1) Relationship between maintenance budget / useful life	1
CG 500	Useful life (standard useful life 80 years)		92	58	96	34	-4	59%	-4%					(cf. PPP School Study2019, margin no. 137) / (2) PPP/BKI more favourable	
CG 500	Residual value risk (maintenance)*		4.887.372 €	-14.212.701 €	6.244.975 €	19.100.073	-1.357.603	-134%	-22%					construction costs lead to hidden reserves (3) Indexing residual value	
CG 500	Additional VAT revenue federal/state/local authorities	1.42%	1.262.331 €												
Life cycle costs excluding risk costs		Nominal	74.114.707 €	68.062.160 €	92.427.044 €	6.052.547 €	-18.312.336 €	9%	-20%						
		NPV	65.483.983 €	61.993.242 €	75.332.603 €	3.490.742 €	-9.848.619 €	6%	-13%						
Life cycle costs including risk costs*		Nominal	69.227.336 €	84.453.842 €	86.182.069 €	-15.226.506 €	-16.954.733 €	-18%	-20%	2,3	3,4	2,6	(A+B+C)/3	PPP incentive structures promote quality management: cost caps, bonus/malus, savings participation, operator liability	1,4
		NPV	63.992.953 €	66.993.987 €	73.427.397 €	-3.001.034 €	-9.434.444 €	-4%	-13%						
Life cycle costs incl. risk costs + additional VAT revenue		Nominal	67.965.004 €	84.453.842 €	86.182.069 €	-16.488.838 €	-18.217.064 €	-20%	-21%						
		NPV	63.992.236 €	66.993.987 €	73.427.397 €	-3.001.751 €	-9.435.181 €	-4%	-13%						

[1] Nominal values indexed over 25 years, base year 2009, Net present value (NPV) - discount rate 5.07 % [2] CG 310 1 = MAX, 2 = PRE Supply costs
[3] Quality assessment 1 = very good / 2 = good / 3 = satisfactory / 4 = sufficient / 5 = unsatisfactory / [4] Questionnaire Construction and operating performance

Economic efficiency study — PPP project no 11 (greenfield/brownfield / education)	Index	PPP[1]	KGSt[1]	BKI[1]	PPP - KGSt	PPP - BKI	PPP/ KGSt	PPP/ BKI	Quality[3] PPP	KGSt	BKI	Value	Remarks	QCO[4]
DIN 276 Construction costs									**2,4**	**3,0**	**3,0**	**100%**	**A. Construction quality, cost and schedule efficiency**	
DIN 277 GFA in m²		16.173	16.173	16.173										
Construction time in months		22	30	30	-8	-8	-26%	-26%	1	3	3	10%	Schedule certainty: 100%	
DIN 276 Construction costs excl. CG 760		9.612.462 €	10.832.696 €	10.832.696 €	-1.220.234	-1.220.234	-11%	-11%	3	3	3	70%	Medium building standard	
Construction costs /sqm GFA in T€		594 €	670 €	670 €	-75	-75	-11%	-11%	1	3	3	10.0%	PPP: Cost certainty: 100%	
CG 760 Interim Financing		254.127 €	335.187 €	335.187 €	-81.060	-81.060	-24%	-24%						
Interest rate		5.170%	4.520%	4.520%	0.65%	0.65%	14%	14%	1	3	3	10.0%	Interest rate hedging promotes high cost and deadline security	
DIN 276 Construction costs incl. CG 760		9.866.589 €	11.167.883 €	11.167.883 €	-1.301.294	-1.301.294	-12%	-12%						
Construction costs /sqm GFA in T€		610 €	691 €	691 €	-80	-80	-12%	-12%						
DIN 18960 Usage Costs excluding risk costs		**45.622.167**	**40.656.414**	**51.048.521**	**4.965.753**	**-5.426.354**	**12%**	**-11%**	**2,6**	**3,3**	**2,8**	**100%**	**B. High-quality maintenance/energy/property management**	
Usage costs including risk costs*		**45.622.167**	**41.508.031**	**51.048.521**	**4.114.135**	**-5.426.354**	**10%**	**-11%**						
CG 100 Capital costs		16.624.086 €	18.634.276 €	18.634.276 €	-2.010.190	-2.010.190	-11%	-11%	3	3	3	10,0%	PPP: Forfaiting with waiver of defence	
CG 100 Capital costs construction costs		14.789.366 €	16.799.556 €	16.799.556 €	-2.010.190	-2.010.190	-12%	-12%						
CG 100 Capital costs IZZB		1.834.720 €	1.834.720 €	1.834.720 €	0	0	0%	0%						
CG 100 Interest financing construction costs		6.040.153 €	6.749.049 €	6.749.049 €	-708.896	-708.896	-11%	-11%						
CG 100 Interest rates		4.847%	4.727%	4.727%	0	0	3%	3%						
CG 200 Property management	1.42%	3.801.272 €	4.470.586 €	3.831.712 €	-669.314	-30.440	-15%	-1%	3	3	3	10.0%	VAT (Personnel costs) / Quality assurance/ Remuneration of	
CG 210 Personnel costs	1.42%	3.729.875 €		3.831.712 €										
CG 210 Own costs of the city	1.42%	71.397 €												
CG 300 Operating costs		17.004.191 €	12.719.896 €	17.639.913 €	4.284.296	-635.722	34%	-4%						
CG 310 Supply costs 2)		3.211.772 €	4.248.521 €	5.734.230 €	-1.036.749	-2.522.458	-24%	-44%	2	3	4	10.0%	Savings of 35% compared to guaranteed max. quantities received by municipality	
CG 311 Water	1.49%	99.544 €	174.306 €	244.828 €	-74.761	-145.284	-43%	-59%					Cost advantage speaks in favour of lower energy consumption	
CG 312 Heating	1.12%	2.432.519 €	2.423.179 €	3.497.215 €	9.340	-1.064.696	0%	-30%					No benchmark comparison of consumption volumes to date	
CG 313 Electricity	3.57%	679.709 €	1.651.037 €	1.992.187 €	-971.328	-1.312.478	-59%	-66%						
CG 320 Waste disposal	1.42%	673.580 €	673.580 €	673.580 €	0	0	0%	0%	3	3	3	10.0%	No qualitative study to date	
CG 330 Cleaning	2.49%	6.494.828 €	6.494.828 €	5.827.629 €	0	667.199	0%	11%	3	3	3	10.0%	No qualitative study to date	
CG 351 Energy management	2.20%	509.238 €			509.238	509.238								
CG 350 Wartung, Inspektion	2.20%	3.415.574 €	494.199 €	3.990.795 €	2.921.375	-575.221	591%	-14%	1	5	1	10.0%	PPP-Anreizstruktur: Service levels und Betreiberhaftung	
CG 370 Taxes and insurance	1.56%		261.802 €	977.321 €	-261.802	-977.321			3	3	3	10.0%	No qualitative study to date	
CG 391 Cafeteria operation	1.42%	393.062 €	393.062 €	393.062 €	0	0	0%	0%	3	3	3	10.0%	No qualitative study to date	
CG 393 Other operating costs	1.42%	2.306.136 €	153.903 €	43.296 €	2.152.233	2.262.840	1398%	5226%	3	3	3	10.0%	PPP may include costs from other cost groups	
CG 400 Repair costs	2.92%	8.192.617 €	5.683.274 €	10.942.620 €	2.509.344	-2.750.002	44%	-25%					Good maintenance management: Service levels, user requests for	
CG 460 Risk of structural damage (maintenance)*	2.92%		851.618 €		-851.618	851.618			2	4	1.5	10.0%	compliance with response/repair times. Cost ceilings: component-specific	
CG 200/350/400 Maintenance		14.008.334 €	6.442.630 €	16.752.458 €	7.565.704	-2.744.123	117%	-16%					maintenance calculation eduction in charges for poor performance;	
CG 200/350/400 in % of restoration costs p.a. / CGST-Target	1.17%	1.56%	0.64%	1.67%									reserve account. CPP: damage risks due to low maintenance budgets	
Revenue / residual value excluding risk costs		**21.127.827 €**	**21.127.827 €**	**21.127.827 €**	**0 €**	**0**	**0%**	**0%**	**2,25**	**4,0**	**2,8**	**100%**	**C. Chance of good residual value without maintenance backlog**	
Revenue / residual value including risk costs		**22.194.889 €**	**17.485.098 €**	**22.380.465 €**	**4.709.791 €**	**-185.576**	**27%**	**-1%**						
Revenue / residual value including risk costs and VAT		**22.903.565 €**	**17.485.098 €**	**22.380.465 €**	**5.418.467 €**	**523.100**	**31%**	**2%**						
CG 500 Residual value	2.92%	22.194.889 €	17.485.098 €	22.380.465 €	4.709.791	-185.576	27%	-1%					(1) Relationship between maintenance budget / useful life	
CG 500 Useful life (standard useful life 80 years)		90	58	92	32	-2	55%	-2%	2,25	4	2	100%	(cf. PPP School Study2019, margin no. 137) / (2) PPP/BKI: more favourable	
CG 500 Residual value risk (maintenance)*		1.067.062 €	-3.642.729 €	1.252.638 €	4.709.791	-185.576	-129%	-15%					construction costs lead to hidden reserves (3) Indexing residual value	
CG 500 Additional VAT revenue federal/state/local authorities	1.42%	708.676 €												
Life cycle costs excluding risk costs (Nominal)		**24.494.340 €**	**19.528.587 €**	**29.920.694 €**	**4.965.753 €**	**-5.426.354 €**	**25%**	**-18%**						
(NPV)		19.383 €	16.620 €	22.562 €	2.763 €	-3.179 €	17%	-14%						
Life cycle costs including risk costs* (Nominal)		**23.427.278 €**	**24.022.933 €**	**28.668.056 €**	**-595.656 €**	**-5.240.778 €**	**-2%**	**-18%**	**2,4**	**3,4**	**2,6**	**(A+B+C)/3**	**PPP incentive structures promote quality management: cost caps, bonus/malus, savings participation, operator liability**	
(NPV)		19.031 €	18.103 €	22.148 €	928 €	-3.117 €	5%	-14%						
Life cycle costs incl. risk costs + additional VAT revenue (Nominal)		**22.718.601 €**	**24.022.933 €**	**28.668.056 €**	**-1.304.332 €**	**-5.949.454 €**	**-5%**	**-21%**						
(NPV)		18.615 €	18.103 €	22.148 €	512 €	3.533 €	3%	-16%						

[1] Nominal values indexed over 25 years, base year 2009 (Nutzungsbeginn). Net present value (NPV) - discount rate: 4,727 % [2] CG 310.1 = MAX, 2 = PRE: Supply costs
[3] Quality assessment: 1 = very good / 2 = good / 3 = satisfactory / 4 = sufficient / 5 = unsatisfactory / [4] Questionnaire Construction and operating performance

Economic efficiency study / PPP project no 12 (greenfield/brownfield / administration)	Index	PPP[1]	KGSt[1]	BKI[1]	PPP - KGSt	PPP - BKI	PPP/ KGSt	PPP/ BKI	Quality[3] PPP	KGSt	BKI	Value	Remarks	QCO[4]
DIN 276 Construction costs									**1,7**	**2,3**	**2,3**	**100%**	**A. Construction quality, cost and schedule efficiency**	
DIN 277 GFA in m2		18.925	18.925	18.925										
Construction time in months		16	26	26	-10	-10	-38%	-38%	1	3	3	10%	Schedule reliability 100%	1
DIN 276 Construction costs excl. CG 760		53.602.008 €	59.598.801 €	59.598.801 €	-5.996.793	-5.996.793	-10%	-10,1%	2	2	2	70%	Upscale building standard	2
Construction costs /sqm GFA in T€		2.832 €	3.149 €	3.149 €	-317	-317	-10%	-10%	1	3	3	10%	PPP. Cost certainty 100%	
CG 760 Interim Financing		-4.451.786 €	1.514.236 €	1.514.238 €	-5.966.023	-5.966.023	-394%	-394%						
Interest rate			3.476%	3.476%					1	3	3	10%		
DIN 276 Construction costs incl. CG 760		49.150.222 €	61.113.038 €	61.113.038 €	-11.962.816	-11.962.816	-20%	-20%						
Construction costs /sqm GFA in T€		2.597 €	3.229 €	3.229 €	-632	-632	-20%	-20%						
DIN 18960 Usage Costs excluding risk costs		**116.525.988**	**122.368.794**	**125.049.652**	**-5.842.806**	**-8.523.664**	**-5%**	**-7%**	**2,6**	**3,4**	**2,9**	**100%**	**B. High-quality maintenance/energy/property management**	
Usage costs including risk costs*		**116.525.988**	**124.671.692**	**125.049.652**	**-8.145.704**	**-8.523.664**	**-7%**	**-7%**						
CG 100 Capital costs		80.766.045 €	98.909.864 €	98.909.864 €	-18.143.819	-18.143.819	-18%	-18%	3	3	3	16,7%	PPP. Forfaiting with waiver of defence	k.A
CG 100 Interest		31.615.823 €	37.796.826 €	37.796.826 €	-6.181.003	-6.181.003	-15%	-16%						
CG 100 Interest rates		4.735%	4.575%	4.575%	0,16%	0,16%	3%	3%						
CG 200 Property management	1,42%	4.113.801 €	3.973.947 €	4.231.388 €	139.854	-117.587	4%	-3%	2	3	3	16,7%	VAT (Personnel costs) / CAFM / Quality assurance/ Remuneration	2
CG 210 Personnel costs	1,42%	3.858.640 €												
CG 210 Own costs of the city	1,42%	255.161 €												
CG 300 Operating costs		16.115.124 €	13.745.869 €	15.846.261 €	2.369.255 €	268.863 €	17%	2%						
CG 310 Supply costs 2)		5.966.786 €	5.598.826 €	5.799.590 €	387.960 €	187.196 €	7%	3%	3	3	3	16,7%	Savings compared to guaranteed maximum quantities are shared	5
CG 311 Water	1,49%	221.316 €	152.678 €	177.628 €	68.638	43.688	45%	25%					Consumption 2013-2019 23% < guaranteed maximum quantity	
CG 312 Heating	1,12%	1.723.461 €	2.287.122 €	2.694.795 €	-563.660	971.334	-25%	-36%					Consumption 2013-2019 32% < guaranteed maximum quantity	
CG 313 Electricity	3,67%	4.042.009 €	3.159.027 €	2.927.167 €	882.982	1.114.842	28%	38%					Consumption 2013-2019 8% > guaranteed maximum quantity (risk company)	
CG 320 Waste disposal	1,42%	559.853 €	559.853 €	559.853 €	0	0	0%	0%	3	3	3	16,7%	No qualitative study to date	
CG 330 Cleaning	2,49%	6.097.439 €	5.575.305 €	3.899.852 €	522.135	2.197.587	9%	56%	3	3	3	16,7%	No qualitative study to date	2
CG 351 Energy management	2,20%													
CG 352 Inspection and maintenance	2,20%	0 €	825.846 €	5.237.914 €	-825.846	-5.237.914			1	5	1	16,7%	PPP incentive structure. High quality due to service levels and operator liability	
CG 370 Taxes and insurance	1,56%	0 €	201.251 €	349.052 €	-201.251	-349.052			3	3	3		See CG 390	
CG 370 Due Diligence	1,56%													
CG 391 Cafeteria operation	1,42%													
CG 393 Other operating costs	1,42%	3.471.045 €	984.788 €	0 €	2.486.257	3.471.045	252%	100%	3	3	3		PPP may include costs from other cost groups	
CG 400 Repair costs	2,92%	15.531.018 €	5.739.113 €	6.062.138 €	9.791.905 €	9.468.879 €	171%	156%					Good maintenance management. Service levels, user requests for	
CG 460 Risk of structural damage (maintenance)*	2,92%		2.302.899 €		-2.302.899	2.302.899							compliance with response/repair times, Cost ceilings, component-specific	
CG 200/350/400 Maintenance	2,92%	19.504.965 €	6.564.959 €	13.063.438 €	12.940.006	6.441.527	197%	49%	2,5	4,5	3,75	16,7%	maintenance calculation eduction in charges for poor performance.	3
CG 200/350/400 in % of restoration costs p.a. / CGST-Target	1,11%	1,29%	0,42%	0,75%									reserve account. CPP: damage risks due to low maintenance budgets	
Revenue / residual value excluding risk costs		**82.040.759 €**	**82.040.759 €**	**82.040.759 €**	**0 €**	**0**	**0%**	**0%**						
Revenue / residual value including risk costs		**83.617.148 €**	**62.178.260 €**	**74.401.336 €**	**21.438.888 €**	**9.215.812**	**34%**	**12%**	**2,5**	**4,5**	**3,8**	**100%**	**C. Chance of good residual value without maintenance backlog**	
Revenue / residual value including risk costs and VAT		**84.296.828 €**	**62.178.260 €**	**74.401.336 €**	**22.118.568 €**	**9.895.492**	**36%**	**13%**						
CG 500 Residual value	2,92%	83.617.148 €	62.178.260 €	74.401.336 €	21.438.888	9.215.812	34%	12%	2,5	4,5	3,75		(1) Relationship between maintenance budget / useful life	3
CG 500 Useful life (standard useful life 80 years)		84	50	65	34	19	68%	29%					(cf. PPP School Study2019, margin no. 137) / (2) PPP/BKI: more favourable	
CG 500 Residual value risk (maintenance)*		1.576.389 €	-19.862.500 €	-7.639.423 €	21.438.888	9.215.812	-108%	-121%					construction costs lead to hidden reserves (3) Indexing residual value	
CG 500 Additional VAT revenue federal/state/local authorities		679.680 €												
Life cycle costs excluding risk costs (Nominal)		**34.485.229 €**	**40.328.034 €**	**43.008.893 €**	**-5.842.806 €**	**-8.523.664 €**	**-14%**	**-20%**						
(NPV)		42.917.207 €	45.968.129 €	48.476.194 €	-3.050.922 €	-5.558.986 €	-7%	-11%						
Life cycle costs including risk costs* (Nominal)		**32.908.840 €**	**62.493.432 €**	**50.648.315 €**	**-29.584.593 €**	**-17.739.476 €**	**-47%**	**-35%**	**2,3**	**3,4**	**3,0**	**(A+B+C)/3**	**PPP incentive structures promote quality management: cost caps, bonus/malus, savings participation, operator liability**	**2,2**
(NPV)		42.328.025 €	54.252.554 €	51.331.465 €	-11.924.529 €	-9.003.440 €	-22%	-18%						
Life cycle costs incl. risk costs + additional VAT revenue (Nominal)		**32.229.160 €**	**62.493.432 €**	**50.648.315 €**	**-30.264.272 €**	**-18.419.156 €**	**-48%**	**-36%**						
(NPV)		42.327.642 €	54.252.554 €	51.331.465 €	-11.924.912 €	-9.003.823 €	-22%	-18%						

[1] Nominal values indexed over 23 years, base year 2012 / Net present value (NPV) - discount rate 4.575 % [2] CG 310 1 = MAX, 2 = PRE. Supply costs
[3] Quality assessment: 1 = very good / 2 = good / 3 = satisfactory / 4 = sufficient / 5 = unsatisfactory / [4] Questionnaire Construction and operating performance

Economic efficiency study — PPP project no 13 (greenfield / education	Index	PPP[1]	KGSt[1]	BKI[1]	PPP - KGSt	PPP - BKI	PPP/ KGSt	PPP/ BKI	Quality PPP	Quality KGSt	Quality BKI	Quality Value	Remarks	QCO[4]
DIN 276 Construction costs									1,4	2,0	2,0	100%	A. Construction quality, cost and schedule efficiency	
DIN 277 GFA in m2		6.995	6.995	6.995										
Construction time in months		13	20	20	-7	-7	-35%	-35%	1	3	3	10%	Schedule reliability 100%	1
DIN 276 Construction costs excl. CG 760		23.021.492 €	24.221.116 €	24.221.116 €	-1.199.624	-1.199.624	-5%	-5.0%	1.5	1.5	1.5	70%	High building standard	2
Construction costs /sqm GFA in T€		3.291 €	3.463 €	3.463 €	-171	-171	-5%	-5%	1	3	3	10%	PPP Cost certainty 100%	
CG 760 Interim Financing		-1.040.195 €	788.494 €	788.494 €	-1.828.688	-1.828.688	-232%	-232%						
Interest rate			3.476%	3.476%					1	3	3	10%	Interest rate hedging promotes high cost and deadline security	
DIN 276 Construction costs incl. CG 760		21.981.298 €	25.009.610 €	25.009.610 €	-3.028.312	-3.028.312	-12%	-12%						
Construction costs /sqm GFA in T€		3.142 €	3.575 €	3.575 €	-433	-433	-12%	-12%						
DIN 18960 Usage Costs excluding risk costs / Usage costs including risk costs*		50.081.222 / 50.081.222	48.868.863 / 49.811.290	54.636.218 / 54.636.218	1.212.359 / 269.931	-4.554.997 / -4.554.997	2% / 1%	-8% / -8%	2,4	3,5	2,8	100%	B. High-quality maintenance/energy/property management	
CG 100 Capital costs		36.018.298 €	40.361.786 €	40.361.786 €	-4.343.488	-4.343.488	-11%	-11%	3	3	3	11,1%	Forfaitierung mit Einredeverzicht	k.A.
CG 100 Interest		14.037.000 €	15.352.176 €	15.352.176 €	-1.315.177	-1.315.177	-9%	-9%						
CG 100 Interest rates		4,705%	4,545%	4,545%	0,16%	0,16%								
CG 200 Property management	1.42%	1.756.162 €	1.518.734 €	1.563.993 €	237.428	192.170	16%	12%	2	3	3	11.1%	VAT (Personnel costs) / CAFM / Quality assurance/ Remuneration	2
CG 210 Personnel costs	1.42%	1.661.851 €											personnel	
CG 210 Own costs of the city	1.42%	94.312 €												
CG 300 Operating costs		4.167.237 €	4.904.769 €	10.939.978 €	737.532 €	6.772.741 €	-15%	-62%						
CG 310 Supply costs 2)	[2]	2.004.836 €	2.001.596 €	2.331.763 €	3.240 €	326.927 €	0%	-14%	2	3	3.5	11.1%	Savings compared to guaranteed maximum quantities are shared	5
CG 311 Water	1,49%	128.997 €	126.082 €	151.265 €	2.915	-22.269	2%	-15%					Consumption 2013-2019: 28% < guaranteed maximum quantity	
CG 312 Heating	1.12%	494.470 €	866.140 €	1.052.389 €	-371.670	-557.920	-43%	-53%					Consumption 2013-2019 22% < guaranteed maximum quantity	
CG 313 Electricity	3.57%	1.381.370 €	1.009.374 €	1.128.108 €	371.996	253.261	37%	22%					Consumption 2013-2019 46% > guar. max. quantity (risk company)	
CG 320 Waste disposal	1.42%	237.370 €	237.370 €	237.370 €	0	0	0%	0%	3	3	3	11.1%	No qualitative study to date	
CG 330 Cleaning	2.49%	1.925.031 €	2.063.177 €	1.687.060 €	-138.147	237.970	-7%	14%	3	3	3	11.1%	No qualitative study to date	2
CG 352 Inspection and maintenance	2.20%		265.689 €	6.476.438 €					1	5	1	11.1%	High-quality maintenance & inspection (see PPP school study 2019)	
CG 370 Taxes and insurance	1,56%		86.228 €	186.793 €	-86.228	-186.793			3	3	3	11.1%		
CG 393 Other operating costs	1.42%		250.709 €	20.553 €	-250.709	-20.553			3	3	3	11.1%		
CG 400 Repair costs	2.92%	8.139.524 €	2.083.574 €	1.770.461 €	6.055.951 €	6.369.063 €	291%	360%					Good maintenance management. Service levels. user requests for	
CG 460 Risk of structural damage (maintenance)*	2.92%		942.427 €		-942.427	942.427							compliance with response/repair times. Cost ceilings, component-specific	
CG 200/350/400 Maintenance	2.92%	9.658.259 €	2.349.263 €	8.246.900 €	7.308.996	1.411.359	311%	17%	2.5	4.5	3.0	11.1%	maintenance calculation eduction in charges for poor performance.	3
CG 200/350/400 in % of restoration costs p.a. / CGST-Target 1.32%		1.38%	0.37%	1.33%									reserve account. CPP damage risks due to low maintenance budgets	
Revenue / residual value excluding risk costs		33.573.971 €	33.573.971 €	33.573.971 €	0 €	0	0%	0%						
Revenue / residual value including risk costs		33.741.223 €	23.037.111 €	33.573.971 €	10.704.112 €	167.252	46%	0%	3,0	4,5	3,0	100%	C. Chance of good residual value without maintenance backlog	
Revenue / residual value including risk costs and VAT		34.033.949 €	23.037.111 €	33.573.971 €	10.996.839 €	459.978	48%	1%						
CG 500 Residual value	2.92%	33.741.223 €	23.037.111 €	33.573.971 €	10.704.112	167.252	46%	0%	3	4.5	3		(1) Relationship between maintenance budget / useful life	2
CG 500 Useful life (standard useful life 80 years)		81	45	80	36	1	80%	1%					(cf. PPP School Study2019, margin no. 137) / (2) PPP/BKI. more favourable	
CG 500 Residual value risk (maintenance)*		167.252 €	-10.536.860 €	0 €	10.704.112	167.252	-102%						construction costs lead to hidden reserves (3) Indexing residual value	
CG 500 Additional VAT revenue federal/state/local authorities	1.42%	292.727 €												
Life cycle costs excluding risk costs (Nominal)	Nominal	16.507.251 €	15.294.892 €	21.062.247 €	1.212.359 €	-4.554.997 €	8%	-22%						
(NPV)	NPV	19.047.773 €	18.074.076 €	21.972.411 €	973.606 €	-2.924.638 €	5%	-13%						
Life cycle costs including risk costs* (Nominal)	Nominal	16.339.999 €	26.774.180 €	21.062.247 €	-10.434.181 €	-4.722.248 €	-39%	-22%	2,2	3,3	2,6	(A+B+C)/3	PPP incentive structures promote quality management; cost caps, bonus/malus, savings participation, operator liability	2,2
(NPV)	NPV	18.984.866 €	22.391.683 €	21.972.411 €	-3.406.817 €	-2.987.545 €	-15%	-14%						
Life cycle costs incl. risk costs + additional VAT revenue (Nominal)	Nominal	16.047.272 €	26.774.180 €	21.062.247 €	-10.726.907 €	-5.014.975 €	-40%	-24%						
(NPV)	NPV	18.984.700 €	22.391.683 €	21.972.411 €	-3.406.983 €	-2.987.711 €	-15%	-14%						

[1] Nominal values indexed over 23 years, base year 2012 / Net present value (NPV) - discount rate 4.545 %   [2] CG 310 1 = MAX, 2 = PRE Supply costs
[3] Quality assessment: 1 = very good / 2 = good / 3 = satisfactory / 4 = sufficient / 5 = unsatisfactory / [4] Questionnaire Construction and operating performance

	Economic efficiency study — PPP project no 14 (greenfield/brownfield / education)	Index	PPP[1]	KGSt[1]	BKI[1]	PPP - KGSt	PPP - BKI	PPP/KGSt	PPP/BKI	PPP	KGSt	BKI	Value	Remarks (Quality[3])	QCO[4]
DIN 276	**Construction costs**									**2,4**	**3,0**	**3,0**	**100%**	**A. Construction quality, cost and schedule efficiency**	
DIN 277	GFA in m²		33.427	33.427	33.427										
	Construction time in months		26 / 2.4 Mio €	26 / 0.4 Mio €	26 / 0.4 Mio €					1	3	3	10%	4 locations in 26 m / construction output p.m. factor 6 / schedule certainty 100%	1
DIN 276	Construction costs excl. CG 760		62.843.832 €	65.530.967 €	65.530.967 €	-2.687.135	-2.687.135	-4%	-4,1%	3	3	3	70%	Medium building standard	1
	Construction costs /sqm GFA in T€		1.880 €	1.960 €	1.960 €	-80	-80	-4%	-4%	1	3	3	10%	PPP: Cost certainty 100%	
CG 760	Interim Financing		1.660.697 €	1.895.972 €	1.895.972 €	-235.275	-235.275	-12%	-12%						
	Interest rate		3.298%	2.698%	2.698%	0.60%	0.60%	22%	22%	1	3	3	10%	Interest rate hedging promotes high cost and deadline certainty	
DIN 276	Construction costs incl. CG 760		64.504.529 €	67.426.939 €	67.426.939 €	-2.922.410	-2.922.410	-4%	-4%						
	Construction costs /sqm GFA in T€		1.930 €	2.017 €	2.017 €	-87	-87	-4%	-4,3%						
DIN 18960	**Usage Costs excluding risk costs**		**168.060.386**	**158.210.233**	**183.328.112**	**9.850.153**	**-15.267.725**	**6%**	**-8%**	**2,5**	**3,3**	**2,8**	**100%**	**B. High-quality maintenance/energy/property management**	
	Usage costs including risk costs		**168.060.386**	**160.156.693**	**183.328.112**	**7.903.693**	**-15.267.725**	**5%**	**-8%**						
CG 100	Capital costs incl. IZZB subsidies		108.994.014 €	111.975.481 €	111.975.481 €	-2.981.467	-2.981.467	-3%	-3%	3	3	3	10,0%	PPP: Forfaiting with waiver of defence	
CG 100	Capital costs IZZB		0 €	0 €	0 €	0	0	0%	0%						
CG 100	Capital costs excl. IZZB subsidies		108.994.014 €	111.975.481 €	111.975.481 €	-2.981.467	-2.981.467	-3%	-3%						
CG 100	Interest		44.489.485 €	44.548.542 €	44.548.542 €	-59.056	-59.056	0%	0%						
CG 100	Interest rates		4.63%	4.46%	4.46%	0.17%	0.17%	4%	4%						
CG 200	Property management	1.42%	15.124.214 €	9.503.595 €	8.202.958 €	5.620.619	6.921.256	59%	84%	2	3	3	10,0%	VAT (Personnel costs) / CAFM / Quality assurance/ Remuneration	3
CG 210	Personnel costs	1.42%	14.267.444 €	9.503.595 €	8.202.958 €										
CG 210	Own costs of the city	1.42%	856.770 €												
CG 300	Operating costs		21.185.528 €	26.152.929 €	38.844.946 €	-4.967.401	-17.659.418	-19%	-45%						
CG 310	Supply costs[2]	1.49%	7.834.460 €	9.152.495 €	12.995.492 €	-1.318.036	-5.161.033	-14%	-40%	2	3	4	10,0%	Consumption/guaranteed volume 2012-2019	1
CG 311	Water		163.114 €	371.068 €	536.241 €	-207.953	-373.126	-56%	-70%					water +43%.	
CG 312	Heating	1.12%	5.006.525 €	5.121.336 €	7.481.787 €	-114.813	-2.475.262	-2%	-33%					heating -61%.	
CG 313	Electricity	3.57%	4.224.689 €	3.660.089 €	4.977.465 €	564.599	-752.776	15%	-15%					electricity: +19%	
CG 320	Waste disposal	1.42%	1.431.901 €	1.431.901 €	1.431.901 €	0	0	0%	0%	3	3	3	10.0%	No qualitative study to date	1
CG 330	Cleaning	2.49%	11.382.996 €	13.097.211 €	12.809.669 €	-1.714.215	-1.426.673	-13%	-11%	3	3	3	10.0%	No qualitative study to date	3
CG 351	Energy management														
CG 350	Inspection and maintenance	2.20%		1.066.880 €	8.709.623 €	-1.066.880	-8.709.623			1	5	1	10.0%	PPP incentive structure: High quality due to service levels and operator liability	
CG 370	Taxes and insurance	1.56%		541.103 €	2.269.401 €	-541.103	-2.269.401			3	3	3		(PPP School Study 2019)	
CG 391	Cafeteria operation		536.172 €	536.172 €	536.172 €	0	0	0%	0%	3	3	3		No qualitative study to date	
CG 393	Other operating costs	1.42%		327.168 €	92.689 €					3	3	3			
CG 400	Repair costs	2.92%	22.756.630 €	12.524.688 €	24.304.726 €	10.231.942 €	-1.548.096	82%	-6%	2	4	2.25	10.0%	Good maintenance management: Service levels, user requests for	1
CG 400	Repair costs	2.92%	22.756.630 €	10.578.228 €	24.304.726 €	12.178.402	-1.548.096	115%	-6%					compliance with response/repair times. Cost ceilings: component-specific	
CG 460	Risk of structural damage (maintenance)*	2.92%		1.946.460 €		-1.946.460	1.946.460							maintenance calculation eduction in charges for poor performance,	
CG 200/350/400	Maintenance		33.343.580 €	15.345.025 €	35.475.236 €	17.998.555	-2.131.657	117%	-6%					reserve account: CPP: damage risks due to low maintenance budgets	
CG 200/350/400	in % of restoration costs p.a. / CGST-Target	#####	1.64%	0.55%	1.51%										
	Revenue / residual value excluding risk costs		**92.518.197 €**	**92.518.197 €**	**92.518.197 €**	**0 €**	**0**	**0%**	**0%**						
	Revenue / residual value including risk costs		**98.396.675 €**	**76.566.784 €**	**97.601.614 €**	**21.829.891 €**	**795.060**	**29%**	**0,8%**	**2,0**	**4,0**	**2,3**	**100%**	**C. Chance of good residual value without maintenance backlog**	
	Revenue / residual value including risk costs and VAT		**100.674.670 €**	**76.566.784 €**	**97.601.614 €**	**24.107.886 €**	**3.073.056**	**31%**	**3%**						
CG 500	Residual value	2.92%	98.396.675 €	76.566.784 €	97.601.614 €	21.829.891	795.060	29%	1%	2	4	2.25		(1) Relationship between maintenance budget / useful life	1
CG 500	Useful life (standard useful life 80 years)		93	58	91									(cf. PPP School Study2019, margin no. 137) / (2) PPP/BKI: more favourable	
CG 500	Residual value risk (maintenance)*		5.878.478 €	-15.951.413 €	6.083.417 €	21.829.891	795.060	-137%	16%					construction costs lead to hidden reserves (3) Indexing residual value	
CG 500	Additional VAT revenue federal/state/local authorities	1.42%	2.277.995 €												
	Life cycle costs excluding risk costs	Nominal	75.542.189 €	65.692.036 €	90.809.915 €	9.850.153 €	-15.267.725 €	15%	-17%						
		NPV	70.374.961 €	64.581.913 €	79.347.073 €	5.793.049 €	-8.972.112 €	9%	-11%						
	Life cycle costs including risk costs*	Nominal	69.663.712 €	83.589.910 €	85.726.497 €	-13.926.198 €	-16.062.786 €	-17%	-19%	**2,3**	**3,4**	**2,7**	(A+B+C)/3	PPP incentive structures promote quality management: cost caps, bonus/malus, savings participation, operator liability	1,4
		NPV	68.312.127 €	70.862.509 €	77.563.236 €	-2.550.383 €	-9.251.109 €	-4%	-12%						
	Life cycle costs incl. risk costs + additional VAT revenue	Nominal	67.385.716 €	83.589.910 €	85.726.497 €	-16.204.193 €	-18.340.781 €	-19%	-21%						
		NPV	66.938.879 €	70.862.509 €	77.563.236 €	-3.923.631 €	-10.624.357 €	-6%	-14%						

[1] Nominal values indexed over 25 years, base year 2011 (Nutzungsbeginn). Net present value (NPV) - discount rate 4.46% [2] CG 310 1 = MAX, 2 = PRE: Supply costs
[3] Quality assessment 1 = very good / 2 = good / 3 = satisfactory / 4 = sufficient / 5 = unsatisfactory / [4] Questionnaire Construction and operating performance

Economic efficiency study PPP project no 15 (greenfield / education sector)	Index	PPP[1]	KGSt[1]	BKI[1]	PPP - KGSt	PPP - BKI	PPP/ KGSt	PPP/ BKI	Quality PPP	Quality KGSt	Quality BKI	Quality Value	Quality Remarks	QCO[4]
DIN 276 Construction costs									**2,4**	**3,0**	**3,0**	**100%**	**A. Construction quality, cost and schedule efficiency**	
DIN 277 GFA in m²		9.369	9.369	9.369										
Construction time in months		17	26	26	-9	-9	-35%	-35%	1	3	3	15%	Cchedule certainty: N.N	2
DIN 276 Construction costs excl. CG 760		17.397.918 €	19.493.190 €	19.493.190 €	-2.095.272 €	-2.095.272 €	-11%	-10.7%	3	3	3	70%	Medium building standard	2
Construction costs /sqm GFA in T€		1 857 €	2 081 €	2 081 €	-224 €	-224 €	-11%	-11%					PPP. Cost certainty: N.N	
CG 760 Interim Financing		369 828 €	488 610 €	488 610 €	-118 782 €	-118 782 €	-24%	-24%						
Interest rate		4 916%	4 316%	4 316%	0.60%	0.60%	14%	14%	1	3	3	15%	Interest rate hedging promotes high cost and deadline security	
DIN 276 Construction costs incl. CG 760		17.767.747 €	19.981.800 €	19.981.800 €	-2.214.053 €	-2 214.053 €	-11%	-11%						
Construction costs /sqm GFA in T€		1 896 €	2 133 €	2 133 €	-236 €	-236 €	-11%	-11.1%						
DIN 18960 Usage Costs excluding risk costs		**44.481.299 €**	**45.243.403 €**	**52.814.848 €**	**-762.104 €**	**-8.333.549 €**	**-2%**	**-16%**	**2,6**	**3,3**	**2,6**	**100%**	**B. High-quality maintenance/energy/property management**	
Usage costs including risk costs*		**44.481.299 €**	**45.723.366 €**	**52.814.848 €**	**-1.242.068 €**	**-8.333.549 €**	**-3%**	**-16%**						
CG 100 Capital costs incl. IZZB subsidies		30.632.418 €	32.951.809 €	32.951.809 €	-2.319.391 €	-2.319.391 €	-7%	-7%	2	3	3	10%	PPP. Forfaiting without waiver of defence (project financing/PFI)	2
CG 100 Capital costs IZZB		4 554 833 €	4 554 833 €	4 554 833 €	0 €	0 €	0%	0%						
CG 100 Capital costs excl. IZZB subsidies		26.077.585 €	28.396.976 €	28.396.976 €	-2.319.391 €	-2.319.391 €	-8%	-8%						
CG 100 Interest		13.919.926 €	13.458.346 €	13.458.346 €	461.580 €	461.580 €	3%	3%						
CG 100 Interest rates		5.301%	4,74%	4,74%	0.56%	0.56%	12%	12%						
CG 200 Property management	1.42%	779.903 €	2.848.310 €	2.052.441 €	-2.068.407	-1.272.537	-73%	-62%	2	3	3	10%	VAT (Personnel costs) / CAFM / Quality assurance/ Remuneration	2
CG 210 Personnel costs	1.42%	712.326 €		2.052.441 €									CG 390 Costs include property management costs	
CG 210 Own costs of the city	1.42%	67.578 €												
CG 300 Operating costs		8.166.557 €	7.031.302 €	9.575.341 €	1.135.255 €	-1.408.784 €	16%	-15%						
CG 310 Supply costs 2)		2 350 867 €	2 947 212 €	2 264 264 €	-596 345 €	86 603 €	-20%	4%	2.5	3	2.25	10%	Savings in consumption volumes go 100% to PPP company no detailed	2
CG 311 Water	1.49%	272.700 €	139.319 €	112.814 €	133.381 €	159.886 €	96%	142%					qualitative study of consumption volumes to date	
CG 312 Heating	1.12%	1.362.989 €	1.742.085 €	1.461.218 €	-379.096 €	-98.229 €	-22%	-7%						
CG 313 Electricity	3.57%	715.179 €	1.065.808 €	690.232 €	-350.630 €	24.946 €	-33%	4%						
CG 320 Waste disposal	1.42%	420.589 €	420.589 €	420.589 €	0 €	0 €	0%	0%	3	3	3	10%	No qualitative study to date	2
CG 330 Cleaning	2.49%	2.401.821 €	2.692.809 €	3.279.430 €	-290.988 €	-877.609 €	-11%	-27%	3	3	3	10%	No qualitative study to date	2
CG 350 Inspection and maintenance	2.20%		309.359 €	2.781.563 €	-309.359 €	-2.781.563 €			2.5	5	1	10%	Service levels/operator liability force intensive inspection&maintenance	
CG 370 Taxes and insurance	1.56%	452.769 €	152.561 €	444.521 €	300.208 €	8.248 €	197%	2%	3	3	3	10%	(PPP-School Study 2019)	
CG 391 Cafeteria operation	1.42%	384.973 €	384.973 €	384.973 €	0 €	0 €	0%	0%	3	3	3	10%	No qualitative study to date	
CG 393 Other operating costs	1.42%	2.155.537 €	123.799 €	0 €	2.031.739 €	2.155.537 €	1641%		2	3	3	10%	Additional costs PFI promote quality management	
CG 400 Repair costs	2.92%	4.902.421 €	2.891.946 €	8.235.258 €	2.010.475 €	-3.332.837 €	70%	-40%					Good maintenance management: Service levels, user requests for	
CG 460 Risk of structural damage (maintenance)*	2.92%		479.963 €		-479.963 €	479.963 €							compliance with response/repair times, Cost ceilings, component-specific	
CG 200/350/400 Maintenance		5 401 049 €	3 397 183 €	12 138 101 €	2 003 866 €	-6 737 052 €	59%	-56%					maintenance calculation eduction in charges for poor performance.	
CG 200/350/400 in % of restoration costs p.a. / CGST-Target 1,07%		1,3%	0,7%	1,9%					2.5	4	1.5	10%	reserve account. CPP. damage risks due to low maintenance budgets	
Revenue / residual value excluding risk costs		**26.747.096 €**	**26.747.096 €**	**26.747.096 €**	**0 €**	**0**	**0%**	**0%**						
Revenue / residual value including risk costs		**27.326.037 €**	**23.707.653 €**	**29.369.360 €**	**3.618.384 €**	**-2.043.323**	**15%**	**-7%**	**2,5**	**4,0**	**1,5**	**100%**	**C Chance auf guten Restwert ohne Instandhaltungsstau**	
Revenue / residual value including risk costs and VAT		**27.439.770 €**	**23.707.653 €**	**29.369.360 €**	**3.732.117 €**	**-1.929.590**	**16%**	**-7%**						
CG 500 Residual value	2.92%	27.326.037 €	23.707.653 €	29.369.360 €	3.618.384 €	-2.043.323 €	15%	-7%	2.5	4	1.5		(1) Relationship between maintenance budget / useful life	2
CG 500 Useful life (standard useful life 80 years)		84	64	102	20	-18	31%	-21%					(cf. PPP School Study2019, margin no. 137) / (2) PPP/BKI: more favourable	
CG 500 Residual value risk (maintenance)*		578 941 €	-3 039 443 €	2 622 264 €	3.618.384 €	-2.043.323 €	-119%	-78%					construction costs lead to hidden reserves (3) Indexing residual value	
CG 500 Additional VAT revenue federal/state/local authorities	1.42%	113.733 €												
Life cycle costs excluding risk costs	Nominal	17.734.203 €	18.496.307 €	26.067.752 €	-762.104 €	-8.333.549 €	-4%	-32%						
	NPV	16.748 €	17.335 €	21.401 €	-587 €	-4.653 €	-3%	-22%						
Life cycle costs including risk costs*	Nominal	17.155.261 €	22.015.713 €	23.445.488 €	-4.860.452 €	-6.290.226 €	-22%	-27%	**2,5**	**3,4**	**2,4**	**(A+B+C)/3**	**PPP incentive structures promote quality management: cost caps, bonus/malus, savings participation, operator liability**	**2,0**
	NPV	16.558 €	18.494 €	20.538 €	-1.936 €	-3.980 €	-10%	-19%						
Life cycle costs incl. risk costs + additional VAT revenue	Nominal	17.041.529 €	22.015.713 €	23.445.488 €	-4.974.184 €	-6.403.959 €	-23%	-27%						
	NPV	16.494 €	18.494 €	20.538 €	-2.001 €	-4.044 €	-11%	-20%						

1) Nominal values indexed over 23.5 years, base year 2008 (begin of usage), Net present value (NPV) - discount rate 4.74 % 2) CG 310.1 = MAX. 2 = PRE. Supply costs
3) Quality assessment: 1 = very good / 2 = good / 3 = satisfactory / 4 = sufficient / 5 = unsatisfactory / 4) Questionnaire Construction and operating performance

Economic efficiency study / PPP project no 16 (greenfield / education sector)		Index	PPP[1]	KGSt[4]	BKI[4]	PPP - KGSt	PPP - BKI	PPP/ KGSt	PPP/ BKI	Quality[3] PPP	KGSt	BKI	Value	Remarks	QCO[4]
DIN 276	**Construction costs**									2,3	2,8	2,8	100%	A. Construction quality, cost and schedule efficiency	
DIN 277	GFA in m²		11.433	11.433	11.433										
	Construction time in months		19	26	26	-7	-7	-27%	-27%	1	3	3	10%	Schedule certainty building: 100%	1
DIN 276	Construction costs excl. CG 760		25.782.069 €	29.476.070 €	29.476.070 €	-3.694.001	-3.694.001	-13%	-13%	2,75	2,75	2,75	70%	Above-average building standard	2
	Construction costs /sqm GFA in T€		2.255 €	2.578 €	2.578 €	-323	-323	-13%	-13%	2	3	3	10%	PPP: Cost certainty +2,5%	
CG 760	Interim Financing		283.303 €	285.261 €	285.261 €	-1.958	-1.958	-1%	-1%						
	Interest rate		1.128%	0.528%	0.528%	0.60%	0.60%	114%	114%	1	3	3	10%	Interest rate hedging promotes high cost and deadline certainty	
DIN 276	Construction costs incl. CG 760		26.065.372 €	29.761.331 €	29.761.331 €	-3.695.959	-3.695.959	-12%	-12%						
	Construction costs /sqm GFA in T€		2.280 €	2.603 €	2.603 €	-323	-323	-12%	-12%						
DIN 1896(	**Usage Costs excluding risk costs**		60.949.336	63.975.403	80.658.774	-3.026.067	-19.709.438	-5%	-24%	2,7	3,3	2,7	100%	B. High-quality maintenance/energy/property management	
	Usage costs including risk costs*		60.949.336	65.347.322	80.658.774	-4.397.986	-19.709.438	-7%	-24%						
CG 100	Capital costs		34.683.369 €	39.601.323 €	39.601.323 €	-4.917.954	-4.917.954	-12%	-12%	3	3	3	10%	Financing of construction costs via local authoritiy	
CG 100	Interest		8.617.997 €	9.839.992 €	9.839.992 €	-1.221.995	-1.221.995	-12%	-12%						
CG 100	Interest rates		2.000%	2.000%	2.000%	0,00%	0.00%								
CG 200	Property management	1.42%	5.347.761 €	4.717.917 €	3.792.172 €	629.844	1.555.589	13%	41%	2	3	3	10%	VAT (Personnel costs) / CAFM / Quality assurance/ Remuneration	1
CG 210	Personnel costs	1.42%	3.339.152 €												
CG 210	Own costs of the city	1.42%	2.008.609 €												
CG 300	Operating costs		15.947.690 €	12.874.296 €	24.435.062 €	3.073.394 € -	8.487.373 €	24%	-35%						
CG 310	Supply costs 2)		2.060.183 €	4.128.502 €	4.604.914 € -	2.068.320 € -	2.544.732 €	-50%	-55%	2	3	3.5	10%	Savings compared to guaranteed maximum quantities are shared	1
CG 311	Water	1.49%	119.568 €	107.847 €	209.723 €	11.720	-90.156	11%	-43%					Cost benefits speak in favour of significantly lower energy consumption	
CG 312	Heating	1.12%	1.055.445 €	2.088.044 €	2.227.097 € -	1.032.600 € -	1.171.652 €	-49%	-53%					No qualitative study to date	
CG 313	Electricity	3.57%	885.170 €	1.932.611 €	2.168.094 €	-1.047.441	-1.282.924	-54%	-59%						
CG 320	Waste disposal	1.42%	756.113 €	756.113 €	756.113 €	0	0	0%	0%	3	3	3	10%	No qualitative study to date	1
CG 330	Cleaning	2.49%	7.908.109 €	6.891.591 €	6.248.878 €	1.016.518	1.659.231	15%	27%	3	3	3	10%	No qualitative study to date	1
CG 351	Energy management	2.20%	174.267 €			174.267	174.267								
CG 352	Inspection and maintenance	2.20%	4.815.691 €	402.445 €	11.366.060 €	4.413.246	-6.550.369	1097%	-58%	2	5	1	10%	PPP incentive structure: High quality due to service levels and operator liability	1
CG 370	Taxes and insurance	1.56%	0 €	255.445 €	1.031.302 €	-255.445	-1.031.302			3	3	3	10%		
CG 370	Due Diligence	1.56%	0 €												
CG 391	Cafeteria operation	1.42%	233.328 €	233.328 €	233.328 €	0	0	0%	0%	3	3	3	10%	No qualitative study to date	
CG 393	Other operating costs	1.42%	0 €	206.872 €	194.467 €	-206.872	-194.467			3	3	3	10%		
CG 400	Repair costs	2.92%	4.970.517 €	6.781.868 €	12.830.217 € -	1.811.351 € -	7.859.700 €	-27%	-61%					Good maintenance management: Service levels, user requests for	1
CG 460	Risk of structural damage (maintenance)*	2.92%		1.371.919 €		-1.371.919	1.371.919							compliance with response/repair times, Cost ceilings, component-specific	
CG 200/350/400	Maintenance	2.92%	14.504.124 €	7.184.313 €	24.196.277 €	7.319.811	-9.692.152	102%	-40%	3	4	1	10%	maintenance calculation eduction in charges for poor performance,	
CG 200/350/400	in % of restoration costs p.a. / CGST-Target	1.20%	1.18%	0.65%	2.04%									reserve account; CPP; damage risks due to low maintenance budgets	
Revenue / residual value excluding risk costs			42.872.478 €	42.872.478 €	42.872.478 €	0 €	0	0%	0%						
Revenue / residual value including risk costs			42.546.865 €	33.115.294 €	48.220.926 €	9.431.571 €	-5.674.062	28%	-12%	3,0	4,0	1,0	100%	C. Chance of good residual value without maintenance backlog	
Revenue / residual value including risk costs and VAT			43.104.636 €	33.115.294 €	48.220.926 €	9.989.342 €	-5.116.290	30%	-11%						
CG 500	Residual value	2.92%	42.546.865 €	33.115.294 €	48.220.926 €	9.431.571	-5.674.062	28%	-12%	3	4	1	100%	(1) Relationship between maintenance budget / useful life	1
CG 500	Useful life (standard useful life 80 years)		79	58	101	21	-22	36%	-22%					(cf. PPP School Study2019, margin no. 137) / (2) PPP/BKI: more favourable	
CG 500	Residual value risk (maintenance)*		-325.614 €	-9.757.185 €	5.348.448 €	9.431.571	-5.674.062	-97%	-106%					construction costs lead to hidden reserves (3) Indexing residual value	
CG 500	Additional VAT revenue federal/state/local authorities	1.42%	557.771 €												
Life cycle costs excluding risk costs		Nominal	18.076.858 €	21.102.925 €	37.786.295 €	-3.026.067 €	-19.709.438 €	-14%	-52%						
		NPV	21.635.872 €	23.962.990 €	36.217.397 €	-2.327.118 €	-14.581.525 €	-10%	-40%						
Life cycle costs including risk costs*		Nominal	18.402.472 €	32.232.029 €	32.437.848 €	-13.829.557 €	-14.035.376 €	-43%	-43%	2,7	3,4	2,2	(A+B+C)/3	PPP incentive structures promote quality management; cost caps, bonus/malus, savings participation, operator liability	1,2
		NPV	21.819.229 €	21.472.120 €	21.375.562 €	347.109 €	443.667 €	2%	2%						
Life cycle costs incl. risk costs + additional VAT revenue		Nominal	17.844.700 €	32.232.029 €	32.437.848 €	-14.387.328 €	-14.593.147 €	-45%	-45%						
		NPV	21.403.246 €	21.472.120 €	21.375.562 €	-68.874 €	27.684 €	0%	0%						

[1] Nominal values indexed over 30 years, base year 2019. Net present value (NPV) - discount rate 2 %   [2] CG 310 1 = MAX, 2 = PRE: Supply costs
[3] Quality assessment: 1 = very good / 2 = good / 3 = satisfactory / 4 = sufficient / 5 = unsatisfactory / [4] Questionnaire Construction and operating performance

Economic efficiency study — PPP project no 17 (brownfield / education sector)		Index	PPP[1]	KGSt[1]	BKI[1]	PPP - KGSt	PPP - BKI	PPP/ KGSt	PPP/ BKI	Quality[3] PPP	KGSt	BKI	Value	Remarks	QCO[4]
DIN 276	**Construction costs**									**2,4**	**3,0**	**3,0**	**100%**	A. Construction quality, cost and schedule efficiency	
DIN 277	GFA in m²		23.300	23.300	23.300										
	Construction time in months		30	39	39	-9	-9	-23%	-23%	1	3	3	10%	Schedule certainty: 100%	
DIN 276	Construction costs excl. CG 760		27.127.029 €	31.559.332 €	31.559.332 €	-4.432.303	-4.432.303	-14%	-14%	3	3	3	70%	Average building standard, heritage conservation project	
	Construction costs /sqm GFA in T€		1.164 €	1.354 €	1.354 €	-190	-190	-14%	-14%	1	3	3	10%	PPP: Cost certainty: +3.5%, remarkably good for a heritage conservation project	
CG 760	Interim Financing		354.111 €	457.579 €	457.579 €	-103.468	-103.468	-23%	-23%						
	interest rate		2.705%	2.105%	2.105%	0.60%	0.60%	29%	29%	1	3	3	10%	Interest rate hedging promotes high cost and deadline certainty	
DIN 276	Construction costs incl. CG 760		27.481.140 €	32.016.911 €	32.016.911 €	-4.535.771	-4.535.771	-14%	-14%						
	Construction costs /sqm GFA in T€		1.179 €	1.374 €	1.374 €	-195	-195	-14%	-14%						
DIN 18960	**Usage Costs excluding risk costs**		**92.553.953**	**82.260.956**	**100.145.441**	**10.292.997**	**-7.591.488**	**13%**	**-8%**	**2,3**	**3,3**	**2,6**	**100%**	B. High-quality maintenance/energy/property management	
	Usage costs including risk costs*		**92.553.953**	**83.578.540**	**100.145.441**	**8.975.413**	**-7.591.488**	**11%**	**-8%**						
CG 100	Capital costs		43.572.779 €	49.723.970 €	49.723.970 €	-6.151.192	-6.151.192	-12%	-12%	3	3	3	11.1%	PPP: Forfaiting with waiver of defence	
CG 100	Interest		16.091.639 €	17.707.059 €	17.707.059 €	-1.615.420	-1.615.420	-9%	-9%						
CG 100	Interest rates		3.46%	3.29%	3.29%	0.17%	0.17%	5%	5%						
CG 200	Property management	1.42%	12.711.809 €	6.503.659 €	6.183.360 €	6.208.150	6.528.448	95%	106%	2	3	3	11.1%	VAT (Personnel costs) / CAFM / Quality assurance/ Remuneration	
CG 210	Personnel costs	1.42%	12.297.577 €		6.183.360 €										
CG 210	Own costs of the city	1.42%	414.232 €												
CG 300	Operating costs		22.128.091 €	18.337.418 €	26.192.320 €	3.790.673	-4.064.229	21%	-16%						
CG 310	Supply costs 2)		4.898.233 €	7.329.885 €	7.330.341 €	-2.431.652	-2.432.108	-33%	-33%	2	3	3	11.1%	Municipality receives 100% of savings compared to guaranteed max. quantities	
CG 311	Water	1.49%	236.596 €	1.083.567 €	1.140.161 €	-846.971	-903.566	-78%	-79%					Consumption operating year 1-15: 32% < guaranteed max. quantity	
CG 312	Heating	1.12%	3.272.234 €	3.678.873 €	4.028.224 €	-406.639	-755.990	-11%	-19%					Consumption operating year 1-15: 43% < guaranteed maximum quantity	
CG 313	Electricity	3.57%	1.389.404 €	2.567.445 €	2.161.956 €	-1.178.042	-772.552	-46%	-36%					Consumption operating year 1-15: 24% > guar. max. quantity; risk company	
CG 320	Waste disposal	1.42%	1.205.219 €	1.205.219 €	1.205.219 €	0	0	0%	0%	3	3	3	11.1%	No qualitative study to date	
CG 330	Cleaning	2.49%	7.461.468 €	8.575.787 €	9.068.994 €	-1.114.319	-1.607.526	-13%	-18%	3	3	3	11.1%	No qualitative study to date	
CG 351	Energy management	2.20%													
CG 350	Inspection and maintenance	2.20%		718.241 €	6.748.351 €	-718.241	-6.748.351			2	5	1	11.1%	ervice level + operator liability force intensive maintenance & inspection	
CG 360	Security and surveillance services			0 €	0 €										
CG 370	Taxes and insurance	1.56%	794.132 €	442.340 €	1.547.381 €	351.792	-753.249	80%	-49%	3	3	3	11.1%	No qualitative study to date	
CG 393	Other operating costs	1.42%	7.769.039 €	65.946 €	292.033 €	7.703.093	7.477.005	11681%	2560%	2	3	3	11.1%	PPP guarantees, administrative costs as part of quality management	
CG 400	Repair costs	2.92%	14.141.275 €	7.695.909 €	18.045.791 €	6.445.366	-3.904.516	84%	-22%					Good maintenance management: Service levels, user requests for	
CG 460	Risk of structural damage (maintenance)*	2.92%		1.317.584 €		-1.317.584	1.317.584							compliance with response/repair times. Cost ceilings, component-specific	
CG 200/350/400	Maintenance	2.92%	16.869.686 €	8.414.149 €	27.921.479 €	8.455.537	-11.051.793	100%	-40%					maintenance calculation eduction in charges for poor performance.	
CG 200/350/400	in % of restoration costs p.a. / CGST-Target　1,19%		1.98%	0.65%	1.81%					1	4	1.5	11.1%	reserve account: CPP: damage risks due to low maintenance budgets	
Revenue / residual value excluding risk costs			**58.877.495 €**	**58.877.495 €**	**58.877.495 €**	**0 €**	**0**	**0%**	**0%**						
Revenue / residual value including risk costs			**64.013.731 €**	**48.726.202 €**	**63.103.152 €**	**15.287.529 €**	**910.579**	**31%**	**1%**	**1,0**	**4,0**	**1,5**	**100%**	C. Chance of good residual value without maintenance backlog	
Revenue / residual value including risk costs and VAT			**65.987.702 €**	**48.726.202 €**	**63.103.152 €**	**17.261.500 €**	**2.884.650**	**35%**	**5%**						
CG 500	Residual value	2.92%	64.013.731 €	48.726.202 €	63.103.152 €	15.287.529	910.579	31%	1%	1	4	1.5	100%	(1) Relationship between maintenance budget / useful life	
CG 500	Useful life (standard useful life 80 years)		99	58	95	41	4	71%	4%					(cf. PPP School Study2019, margin no. 137) / (2) PPP/BKI: more favourable	
CG 500	Residual value risk (maintenance)*		5.136.237 €	-10.151.292 €	4.225.658 €	15.287.529	910.579	-151%	22%					construction costs lead to hidden reserves (3) Indexing residual value	
CG 500	Additional VAT revenue federal/state/local authorities	1.42%	1.973.971 €												
Life cycle costs excluding risk costs		Nominal	**33.676.458 €**	**23.383.461 €**	**41.267.946 €**	**10.292.997 €**	**-7.591.488 €**	**44%**	**-18%**						
		NPV	**37.537 €**	**29.411 €**	**40.554 €**	**8.126 €**	**-3.017 €**	**28%**	**-7%**						
Life cycle costs including risk costs*		Nominal	**28.540.221 €**	**34.852.338 €**	**37.042.289 €**	**-6.312.116 €**	**-8.502.067 €**	**-18%**	**-23%**	**1,9**	**3,4**	**2,4**	**(A+B+C)/3**	PPP incentive structures promote quality management: cost caps, bonus/malus, savings participation, operator liability	
		NPV	**35.462 €**	**34.044 €**	**38.847 €**	**1.418 €**	**-3.385 €**	**4%**	**-9%**						
Life cycle costs incl. risk costs + additional VAT revenue		Nominal	**26.566.251 €**	**34.852.338 €**	**37.042.289 €**	**-8.286.087 €**	**-10.476.038 €**	**-24%**	**-28%**						
		NPV	**34.198 €**	**34.044 €**	**38.847 €**	**153 €**	**-4.650 €**	**0%**	**-12%**						

[1] Nominal values indexed over 29 years, base year 2005. Net present value (NPV) - discount rate 4.74 %　[2] CG 310: 1 = MAX, 2 = PRE. Supply costs

[3] Quality assessment: 1 = very good / 2 = good / 3 = satisfactory / 4 = sufficient / 5 = unsatisfactory / [4] Questionnaire Construction and operating performance

Economic efficiency study PPP project no 18 (brownfield / education sector)		Index	PPP[1]	KGSt[1]	BKI[1]	PPP - KGSt	PPP - BKI	PPP/ KGSt	PPP/ BKI	Quality[3] PPP	KGSt	BKI	Value	Remarks	QCO[4]
DIN 276	**Construction costs**									**2,4**	**3,0**	**3,0**	**100%**	A. Construction quality, cost and schedule efficiency	
DIN 277	GFA in m²		7.691	7.691	7.691										
	Construction time in months		17	26	26	-9	-9	-35%	-35%	1	3	3	15%	Schedule certainty: N.N	2
DIN 276	Construction costs excl. CG 760		9.000.256 €	9.824.364 €	9.824.364 €	-824.108	-824.108	-8%	-8%	3	3	3	70%	Medium building standard	2
	Construction costs /sqm GFA in T€		1 170 €	1 277 €	1 277 €	-107	-107	-8%	-8%					PPP: Cost certainty: N.N	
CG 760	Interim Financing		359 811 €	415 913 €	415 913 €	-56 102	-56 102	-13%	-13%						
	Interest rate		4.973%	4.373%	4.373%	0.60%	0.60%	14%	14%	1	3	3	15%	Interest rate hedging promotes high cost and deadline certainty	
DIN 276	Construction costs incl. CG 760		9.360.067 €	10.240.277 €	10.240.277 €	-880.210	-880.210	-9%	-9%						
	Construction costs /sqm GFA in T€		1 217 €	1 331 €	1 331 €	-114	-114	-9%	-9%						
DIN 18960	**Usage Costs excluding risk costs** **Usage costs including risk costs***		30.017.790 30.017.790	27.265.058 27.665.727	32.139.972 32.139.972	2.752.732 2.352.063	-2.122.182 -2.122.182	10% 9%	-7% -7%	**2,7**	**3,3**	**2,7**	**100%**	B. High-quality maintenance/energy/property management	
CG 100	Capital costs		16.415.890 €	17.026.836 €	17.026.836 €	-610.947	-610.947	-4%	-4%	2.0	3	3	10%	PPP: Forfaiting with waiver of defence	2
CG 100	Interest		7 055 823 €	6 786 559 €	6 786 559 €	269 263	269 263	4%	4%						
CG 100	Interest rates		5.3%	4.74%	4.74%	0.56%	0.56%	12%	12%						
CG 200	Property management	1.42%	846.425 €	2.066.998 €	1.784.114 €	-1.220.572	-937.689	-59%	-53%	2.0	3	3	10%	VAT (Personnel costs) / CAFM / Quality assurance/ Remuneration	2
CG 210	Personnel costs	1.42%	775.028 €		1.784 114 €									CG 390 Costs include property management costs (PFI)	
CG 210	Own costs of the city	1.42%	71.397 €												
CG 300	Operating costs		8.779.495 €	6.002.131 €	8.345.275 €	2.777.364	434.220	46%	5%						2
CG 310	Supply costs 2)		3 004 996 €	1 939 396 €	2 642 528 €	1 065 600	362 468	55%	14%	4.0	3	4	10%	Savings in consumption volumes go 100% to PPP company	
CG 311	Water	1.49%	170 411 €	80 477 €	122 396 €	89 935	48 016	112%	39%						
CG 312	Heating	1.12%	2 294 523 €	1 126 903 €	1 682 375 €	1 167 620	612 148	104%	36%						
CG 313	Electricity	3.57%	540 062 €	732 016 €	837 758 €	-191 954	-297 696	-26%	-36%						
CG 320	Waste disposal	1.42%	311 433 €	311 433 €	311 433 €	0	0	0%	0%	3.0	3	3	10%	No qualitative study to date	2
CG 330	Cleaning	2.49%	2 416 393 €	2 940 144 €	2 670 796 €	-523 751	-254 403	-18%	-10%	3.0	3	3	10%	No qualitative study to date	
CG 351	Energy management	2.20%													
CG 350	Inspection and maintenance	2.20%		225 003 €	1 836 842 €	-225.003	-1 836 842			2.5	5	1	10%	Service levels/operator liability force intensive inspection and maintenance	
CG 360	Security and surveillance services													(PPP School Study 2019)	
CG 370	Taxes and insurance	1.56%	477 986 €	108 264 €	454 063 €	369 722	23 923	342%	5%	3.0	3	3	10%	No qualitative study to date	
CG 391	Cafeteria operation	1.42%	406 734 €	406 734 €	406 734 €	0	0	0%	0%	3.0	3	3	10%	No qualitative study to date	
CG 393	Other operating costs	1.42%	2 161 953 €	71 158 €	22 880 €	2 090 795	2 139 073	2938%	9349%	2.0	3	3	10%	Administrative costs, additional costs PFI promote quality management	
CG 400	Repair costs	2.92%	3.975.980 €	2.569.762 €	4.983.747 €	1.406.218	-1.007.767	55%	-20%					Good maintenance management: Service levels, user requests for	
CG 460	Risk of structural damage (maintenance)"	2.92%		400 669 €		-400.669	400 669							compliance with response/repair times, Cost ceilings, component-specific	
CG 200/350/400	Maintenance		4 568 478 €	2 877 440 €	7 625 763 €	1 691 038	-3 057 285	59%	-40%					maintenance calculation eduction in charges for poor performance,	2
CG 200/350/400 in % of restoration costs p.a. / CGST-Target		1.17%	1.53%	0.61%	1.63%					2.5	4	1.5	10%	reserve account, CPP: damage risks due to low maintenance budgets	
	Revenue / residual value excluding risk costs **Revenue / residual value including risk costs** **Revenue / residual value including risk costs and VAT**		19.044.400 € 19.919.781 € 20.036.968 €	19.044.400 € 15.551.408 € 15.551.408 €	19.044.400 € 20.090.796 € 20.090.796 €	0 € 4.368.373 € 4.485.559 €	0 -171.015 -53.828	0% 28% 29%	0% -1% 0%	**2,5**	**4,0**	**2,0**	**100%**	C. Chance of good residual value without maintenance backlog	
CG 500	Residual value	2.92%	19.919.781 €	15.551.408 €	20.090.796 €	4.368.373	-171.015	28%	-1%	2.5	4	2		(1) Relationship between maintenance budget / useful life	2
CG 500	Useful life (standard useful life 80 years)		89	57	91	32	-2	56%	-2%					(cf. PPP School Study2019, margin no. 137) / (2) PPP/BKI: more favourable	
CG 500	Residual value risk (maintenance)"		875 381 €	-3 492 992 €	1 046 396 €	4 368 373	-171 015	-125%	-16%					construction costs lead to hidden reserves (3) Indexing residual value	
CG 500	Additional VAT revenue federal/state/local authorities	1.42%	117.186 €												
	Life cycle costs excluding risk costs	Nominal NPV	10.973.390 € 11.146 €	8.220.658 € 9.561 €	13.095.572 € 12.347 €	2.752.732 € 1.584 €	-2.122.182 € -1.202 €	33% 17%	-16% -10%						
	Life cycle costs including risk costs*	Nominal NPV	10.098.009 € 10.858 €	12.114.319 € 10.843 €	12.049.176 € 12.603 €	-2.016.310 € 15 €	-1.951.167 € -1.145 €	-17% 0%	-16% -10%	**2,5**	**3,4**	**2,6**	**(A+B+C)/3**	PPP incentive structures promote quality management: cost caps, bonus/malus, savings participation, operator liability	**2,0**
	Life cycle costs incl. risk costs + additional VAT revenue	Nominal NPV	9.980.823 € 10.790 €	12.114.319 € 10.843 €	12.049.176 € 12.603 €	-2.133.497 € -52 €	-2.068.354 € -1.213 €	-18% 0%	-17% -10%						

[1] Nominal values indexed over 25 years, base year 2008 (begin of usage). Net present value (NPV) - discount rate: 4.74 %. [2] CG 310.1 = MAX, 2 = PRE. Supply costs
[3] Quality assessment: 1 = very good / 2 = good / 3 = satisfactory / 4 = sufficient / 5 = unsatisfactory / [4] Questionnaire Construction and operating performance

ANNEX B1 — Internet research — Delays in school construction projects

lfd. Nr.	Überschrift	Standort	Quelle	Datum	Verzögerung	Fundstelle
1	Fertigstellung der Rickenbacher Schule verzögert sich	Rickenbach	Badische Zeitung	20.02.2020	6 Monate	https://www.badische-zeitung.de/fertigstellung-der-rickenbacher-schule-verzoegert-sich--182950013.html
2	Fertigstellung der Otto-Hahn-Schulen erneut verschoben	Bergisch Gladbach	IGL Bürgerportal	31.10.2019	12 Monate	https://in-gl.de/2019/10/31/fertigstellung-der-otto-hahn-schulen-erneut-verschoben/
3	Neubau der Wiehagenschule: Verantwortliche räumen Fehler in öffentlicher Aufklärung ein	Werne	Ruhr Nachrichten	22.06.2019	3-4 Monate	https://www.ruhrnachrichten.de/werne/wiehagenschule-neubau-fehler-stadtverwaltung-eingeraeumt-werne-1419388.html
4	Teilmodernisierung der Sporthalle Geschwister Scholl Schule	Leipzig	Kleine Anfrage Freibeuterfraktion	23.07.2017	2 Monate	https://fdp-stadtrat-leipzig.de/anfragen/verzoegerungen-bei-bau-und-sanierungsarbeiten-an-schulen-und-kitas/
5	Brandschutzmaßnahme Forderschule Rosenweg	Leipzig	Kleine Anfrage Freibeuterfraktion	23.07.2017	8 Monate	https://fdp-stadtrat-leipzig.de/anfragen/verzoegerungen-bei-bau-und-sanierungsarbeiten-an-schulen-und-kitas/
6	Energetische Sanierung F.A.Brockhaus-Schule	Leipzig	Kleine Anfrage Freibeuterfraktion	23.07.2017	12 Monate	https://fdp-stadtrat-leipzig.de/anfragen/verzoegerungen-bei-bau-und-sanierungsarbeiten-an-schulen-und-kitas/
7	Neue Grundschule: Fertigstellung verzögert sich weiter	Peine/Stederdorf	Peiner Allgemeine	22.02.2018	6 Monate +x	https://www.paz-online.de/Stadt-Peine/Neue-Schule-in-Stederdorf-Fertigstellung-verzoegert-sich-weiter
8	Firma spurlos verschwunden: Bauverzögerung an der Grundschule in Rickenbach	Rickenbach	Südkurier	24.09.2020	+ 3 Monate	https://www.suedkurier.de/region/hochrhein/rickenbach/firma-spurlos-verschwunden-bauverzoegerung-an-der-grundschule-in-rickenbach;art372616,10622324
9	Lindenschule Fertigstellung verzögert sich	Stadt Frechen	Rheinische Anzeigenblatter	22.03.2013	+ 5 Monate	https://www.rheinische-anzeigenblaetter.de/mein-blatt/wochenende/frechen/lindenschule-fertigstellung-verzoegert-sich-29910618
10	Darum verzögert sich die Fertigstellung des Schulzentrums Aspe um Jahre	Bad Salzuflen / Lippe	LZ.DE	10.09.2020	+ 36 Monate	https://www.lz.de/lippe/bad_salzuflen/22858394_Darum-verzoegert-sich-die-Fertigstellung-des-Schulzentrums-Aspe-um-Jahre.html
11	Von oben bis unten durchnässt: Wasserschaden verzögert Schuleröffnug	Gemeinde Haar	Merkur.de	03.12.2020	+ 3 Monate	https://www.radiolippe.de/nachrichten/lippe/detailansicht/verzoegerungen-am-schulzentrum-aspe-werden-teuer.htm?L=0&cHash=fb6115c15c2b6d4d574671c8c17d7886
12	Schulzentrum wegen Cyberangriff später fertig	Potsdam	Potsdamer neueste Nachrichten	28.06.2020	+ 6 Monate	https://www.merkur.de/lokales/muenchen-lk/haar-ort104496/von-oben-bis-unten-durchnaesst-90045570.html
13	Fertigstellung der Paul-Gerhardt-Schule verzögert sich - Umzug in den Herbstferien	Euskirchen	Homepage der Stadt	14.08.2019	+ 3 Monate	https://www.euskirchen.de/leben-in-euskirchen/aktuelle-mitteilungen/detail/news/2019/8/14/fertigstellung-der-paul-gerhardt-schule-verzoegert-sich-umzug-in-den-herbst
14	Umzug im Juli 2020 - Grundschule an der Karl-Sittler-Straße wird 1 Jahr später fertig	Poing	Süddeutsche Zeitung	17.12.2019	+12 Monate	https://www.sueddeutsche.de/muenchen/ebersberg/schulgebaeude-in-poing-umzug-im-juli-2020-1.4727037
15	Richtfest verschiebt sich um ein Jahr	Freiberg	Bietigheimer Zeitung	19.06.2020	+12 Monate	https://www.bietigheimerzeitung.de/inhalt.freiberg-richtfest-verschiebt-sich-um-ein-jahr.e39744b4-971d-43e7-b7dc-f916c9a01a42.html
16	Verzögerung bei Fertigstellung der Theodor-Fontane-Schule	Ludwigsfelde	Homepage der Stadt	27.07.2017	+ 3 Monate	https://www.ludwigsfelde.de/verzoegerung-bei-fertigstellung-der-theodor-fontane-schule/
17	Schwerin: Erich-Weinert-Schule bleibt Baustelle	Schwerin	Schwerin lokal	03.12.2020	+ 6 Monate	https://schwerin-lokal.de/schwerin-erich-weinert-schule-bleibt-baustelle/
18	Umzug der Mühlbergschule in Frankfurt verzögert sich	Frankfurt Sachsenhausen	Frankfurter Rundschau	27.09.2020	+ 3 Monate + x?	https://www.fr.de/frankfurt/sachsenhausen-ort29377/umzug-der-muehlbergschule-in-frankfurt-verzoegert-sich-90054533.html
19	Fertigstellung der Max-von-Laue-Schule verzögert sich erneut	Berlin - Lichterfelde	Berliner Woche	19.08.2013	+ 3 Monate	https://www.berliner-woche.de/lichterfelde/c-sonstiges/fertigstellung-der-max-von-laue-schule-verzoegert-sich-erneut_a33565
20	Nachsitzen beim Magdeburger Schulbau (Verzögerungen bei 8 Schulbauprojekten)	Magdeburg	www.volksstimme.de	25.06.2020	+ 1-3 Monate +x	https://www.volksstimme.de/lokal/magdeburg/verzug-nachsitzen-beim-magdeburger-schulbau
21	Frankfurt: Umzug der Merianschule verzögert sich schon wieder - Elternschaft verärgert: „Das ist ein Witz"	Frankfurt	Frankfurter Rundschau	08.09.2020	24 Monate	https://www.fr.de/frankfurt/nordend-ort904333/nordend-merianschule-muss-weiter-warten-90038955.html
22	Neubau Schollschule: Fertigstellung verzögert sich	Jüterbog	Märkische Allgemeine	16.02.2020	+ 2 Monate	https://www.maz-online.de/lokales/Teltow-Flaeming/Jueterbog/Neubau-Schollschule-Jueterbog-Bauarbeiten-dauern-laenger
23	Einweihungsparty erst im September 2019 - Verzögerung um 5 Monate	Elmshorn	Elmshorner Nachrichten	15.12.2018	+ 6 Monate	https://www.shz.de/lokales/elmshorner-nachrichten/einweihungsparty-erst-im-september-2019-id21983002.html
24	Fertigstellung des Schulneubaus in Kolkwitz/Gołkojce verzögert sich	Landkreis Spree-Neiße	Homepage	07.05.2020		https://www.lkspr.de/aktuelles/aktuelles-landkreis-spree-neisse/pressearchiv/27025-fertigstellung-des-schulneubaus-in-kolkwitzgokojce-verzogert-sich.html
25	Kreuzschule: Der Bau verzögert sich	Regensburg	Mittelbayerische Nachrichten	23.07.2019	+ 3 Monate	https://www.mittelbayerische.de/region/regensburg-stadt-nachrichten/kreuzschule-der-bau-verzoegert-sich-21179-art1808576.html
26	Sanierung der Regionalen Schule "Erich Weinert" verzögert sich!	Schwerin	Homepage Schule	01.10.2020		https://www.weinertschule-schwerin.de/Aktuelles/Schulsanierung
27	Stadt München: Ein Jahr Verzögerung für neue Grundschule an der Theodor-Fischer-Straße - BA verärgert	München-Untermenzing	Hallo München	19.03.2020	+12 Monate	https://www.hallo-muenchen.de/muenchen/west/muenchen-untermenzing-grundschule-theodor-fischer-verzoegerung-neubau-13606317.html
28	Brandschutz: Behelfsschule verzögert sich	Lübeck	HL-Live	03.12.2020	+ 6 Monate	https://www.hl-live.de/text.php?id=134885
29	Neubau der Paul-Winter-Schule verzögert sich um mehrere Monate - Schuldfrage ist offen	Neuburg	Donaukurier	06.12.2018	mehrere Monate	https://www.donaukurier.de/lokales/neuburg/Falscher-Bodenaustausch-auf-Baustelle;art1763,4009668
30	Keine Klage riskieren: Neubau des Gutenberg-Gymnasiums verzögert sich	Mainz	Allgemeine Zeitung	15.04.2019	einige Monate	https://www.allgemeine-zeitung.de/lokales/mainz/nachrichten-mainz/keine-klage-riskieren-neubau-des-gutenberg-gymnasiums-verzoegert-sich_20085420
31	Handwerkermangel verzögert Kita-Neubau in Dollbergen	Uetze	Hannoversche Allgemeine	15.04.2019	+ 3 / mehrere Monate	https://www.haz.de/Umland/Uetze/Uetze-Handwerkermangel-verzoegert-Kita-Neubau-in-Dollbergen
32	Alles andere als ideal - Neubau Grundschule Grandlstraße verzögert sich	München-Pasing	Wochenanzeiger München	25.04.2017	+ 3 Monate	https://www.wochenanzeiger-muenchen.de/allach-menzing/alles-andere-als-ideal,90386.html
33	MPG-Neubau in Groß-Umstadt verzögert sich	Groß-Umstadt	Echo-online	10.03.2020	Arbeiten hinter Zeitplan	https://www.echo-online.de/lokales/darmstadt-dieburg/gross-umstadt/mpg-neubau-in-gross-umstadt-verzoegert-sich_21389494
34	Wasserschaden: OGS-Neubau der Overbergschule wird erst im April fertig	Lünen	Ruhrnachrichten	20.11.2020	+ 3 Monate	https://www.ruhrnachrichten.de/luenen/wasserschaden-ogs-neubau-der-overbergschule-wird-erst-im-april-fertig-1576241.html
35	Mensabau für Schwanenschule verzögert sich	Wermelskirchen	RP-online	30.10.2019	+ 3 Monate	https://rp-online.de/nrw/staedte/wermelskirchen/wermelskirchen-mensabau-fuer-schwanenschule-verzoegert-sich_aid-46823647
36	Verzögerungen kosten die Stadt Millionen: Bürgermeister hat Plan zum Schul-Schnellbau	Dresden	Tag24	01.09.2020	viele Vorhaben 1 Jahr	https://www.tag24.de/dresden/politik-wirtschaft/verzoegerungen-kosten-die-stadt-millionen-buergermeister-jan-donhauser-hat-einen-plan-zum-schul-schnell-bau-stesad-1
37	Riesiges Schulbauprojekt gerät ins Stocken	Kreis Ludwigsburg	Stuttgarter Zeitung	09.03.2020	+ 6 Monate	https://www.stuttgarter-zeitung.de/inhalt.kreis-ludwigsburg-riesiges-schulbauprojekt-geraet-ins-stocken.7cc3422f-1820-4232-b2a6-b88b26de95a3.html
38	Neubau der John Cranko Schule wird erst 2019 eröffnet	Stuttgart	Beteiligungsportal BW	17.10.2017	+12 Monate	https://beteiligungsportal.baden-wuerttemberg.de/de/informieren/service/pressemitteilung/pid/neubau-der-john-cranko-schule-wird-erst-2019-eroeffnet/
39	Neubau an der Ihringshäuser Grundschule verzögert sich	Fuldatal	HNA.de	18.07.2017	+ 3 Monate	https://www.hna.de/lokales/kreis-kassel/fuldatal-ort83863/neubau-an-ihringshaeuser-grundschule-verzoegert-sich-8493814.html
40	Der Neubau der Schule an der Nathrather Straße verzögert sich um etwa sechs Monate	Wuppertal	Westdeutsche Zeitung	28.07.2018	+ 6 Monate ggf. + 9	https://www.wz.de/nrw/wuppertal/grundschul-rohbau-fast-fertig_aid-25009629
41	Bauboom wirbelt Pläne durcheinander	Achim	kreiszeitung.de	30.10.2019	+ 3 Monate	https://www.kreiszeitung.de/lokales/verden/achim-ort44553/bauboom-wirbelt-plaene-durcheinander-13122115.html
42	Egmating: Umzug ins neue Rathaus verschoben	Egmating	Meine Anzeigenzeitung	19.09.2019	+ 6 Monate	https://www.meine-anzeigenzeitung.de/lokales/ebersberg/umzug-egmating-verzoegert-sich-auch-konsequenzen-grundschueler-13018558.html
43	Verzögerung beim Neubau der Mathilde-Anneke-Schule	Münster	Homepage der Stadt	01.06.2018	+12 Monate	https://www.wn.de/Muenster/3324468-Verzoegerung-beim-Neubau-der-Mathilde-Anneke-Schule-Schule-wird-ein-Jahr-spaeter-fertig
44	Fertigstellung Berufskolleg-Campus Moers verzögert sich	Moers	Kreis Wesel Homepage	31.08.2018	+12 Monate	https://www.kreis-wesel.de/de/presse/fertigstellung-berufskolleg-campus-moers-verzoegert-sich/
45	Grundschule: Klage verzögert die Fertigstellung des Neubaus um ein halbes Jahr	Frankfurt Berkersheim	Frankfurter Neue Presse	29.08.2018	+ 6 Monate	https://www.fnp.de/lokales/grundschule-klage-verzoegert-fertigstellung-neubaus-halbes-jahr-10370958.html
46	"Belastetes Material" unter der Goetheschule: Neubau verzögert sich und wird teurer	Wetzlar	mittelhessen.de	25.06.2019	+ 2,5 Monate	https://www.mittelhessen.de/lokales/wetzlar/wetzlar/belastetes-material-unter-der-goetheschule-neubau-verzoegert-sich-und-wird-teurer_20238591
47	Fertigstellung der Bildungslandschaft Altstadt Nord verzögert sich	Köln	Homepage der Stadt	13.03.2020	neuer Zeitplan i.A.	https://www.stadt-koeln.de/politik-und-verwaltung/presse/mitteilungen/21543/index.html
48	Heide-Ost: Eröffnung der Turnhallen verzögert sich um Monate	Heide	Boyens Medien	02.08.2019	+ 4 Monate	https://www.boyens-medien.de/artikel/dithmarschen/heide-ost-eroeffnung-der-turnhallen-verzoegert-sich-um-monate-285477.html?NULL

Durchschnittliche Verzögerung — ? Monate

ANNEX B2 — Internet research — Explosion in costs for school buildings

lfd. Nr.	Überschrift	Standort	Quelle	Datum	Baukosten in Mio.€			Ursachen	Neubau/ Sanierung	Fundstelle
					SOLL	IST	Diff.			
1	Kostenexplosion bei Schulbauten; Berufsschulzentrum Nord	Darmstadt	Echo-online	22.03.2019	115,5	126,7	10%	erheblich gestiegende Vergaberisiken, nur 1 Angebot San+Erweiterung		https://www.echo-online.de/lokales/darmstadt/kostenexplosion-bei-schulbauten_20031652
2	Kostenexplosion bei Schulbauten: Neubau Grundschule+Kita Lincoln-Siedlung	Darmstadt	Echo-online	23.03.2019	27,9	33,8	21%	erheblich gestiegende Vergaberisiken, nur 1 Angebot Neubau		https://www.echo-online.de/lokales/darmstadt/kostenexplosion-bei-schulbauten_20031653
3	Kostenexplosion bei Schulbauten	Bayern	br.de	26.11.2017				3 Beispiele stellvertretend für die GEsamtlage in Bayern; kommunale Forderung; Erh...		https://www.br.de/nachrichten/bayern/kostenexplosion-bei-schulbauten,Qc1775b
4	Kostenexplosion bei Schulbau Theodor Heuß Gymnasium	LK Donau-Ries	br.de	27.11.2017	20,0	30,0	50%	keine Preissteigerung bei Kostenplanung, unliebsame Überraschungen im Bauverlau...		https://www.br.de/nachrichten/bayern/kostenexplosion-bei-schulbauten,Qc1775b
5	Kostenexplosion bei Schulzentrum	Rain am Lech	br.de	28.11.2017	25,0	60,0	140%			https://www.br.de/nachrichten/bayern/kostenexplosion-bei-schulbauten,Qc1775b
6	Kostenexplosion bei Berufsschule Vilshofen	Vilshofen	br.de	29.11.2017	35,0	70,0	100%			https://www.br.de/nachrichten/bayern/kostenexplosion-bei-schulbauten,Qc1775b
7	Grundschule wird eine Million Euro teurer	Schongau	Merkur	26.11.2015	17,2	20,0	16%	Zusätzliche Abbruchkosten; Erhöhung Bodenplatte wg. Grundwasserproblemer etc.		https://www.merkur.de/lokales/schongau/neue-kostenberechnung-neue-grundschule-wird-eine-m...
8	Kostenexplosion bei Schulbau	Lahr	Badische Zeitung	20.07.2007		0,9				https://www.merkur.de/lokales/schongau/neue-kostenberechnung-neue-grundschule-wird-m...
9	Kostenexplosion bei bei Errichtung der dritten Gesamtschule	Gütersloh	UWG-Gütersloh	13.11.2019			70%			https://www.uwg-guetersloh.de/index.php/2019/11/13/hier-ist-nichts-schiefgelaufen-kostenexplos...
10	Schulbaukosten in Frankfurt explodieren - Revisionsamt offenbart Planungsfehler	Frankfurt	FNP.de	05.12.2018				Kostenexplosionen sind in Frankfurt mittlerweile eher die Regel als die Ausnahme.		https://www.fnp.de/frankfurt/schulbaukosten-frankfurt-explodieren-10804404.html
11	- KGS Niederrad	Frankfurt	FNP.de	06.12.2018	22,4	29,5	32%			https://www.fnp.de/frankfurt/schulbaukosten-frankfurt-explodieren-10804404.html
12	- IGS Kalbach-Riedberg	Frankfurt	FNP.de	07.12.2018	39,6	46,9	18%			https://www.fnp.de/frankfurt/schulbaukosten-frankfurt-explodieren-10804404.html
13	Sophie-Opel-Schule-NEUBAU IN RÜSSELSHEIM CDU will „die Notbremse ziehen"	Rüsselsheim	FNP.de	08.11.2016				Anstieg Planungskosten von 30,5 Mio. auf über 50 Mio. €		https://www.fnp.de/lokales/kreis-gross-gerau/ruesselsheim-ort29367/viel-darf-schule-kosten-1054...
14	Stadt ringt beim Schulbau mit Kostenexplosionen	Dresden	Pressreader	13.08.2019						https://www.google.com/search?q=Kostenexplosion+bei+SChulbau&ei=2R_KX47IIMK8kwXQ1SD4A
15	Stadt Köln: Verdopplung der Kostensteigerung bei städtischen Großbauprojekten	Köln	die Wirtschaft-Koeln	18.07.2018				15% Durchschr. 15,46% (2018); Entwicklung Kostensteigerung b. Großbauprojekten; 2017		https://www.diewirtschaft-koeln.de/kostensteigerung-bei-koelner-bauprojekten-_id3927.html
16	Kölner Bildungs-Campus Kosten für Schulbauten am Klingelpützpark explodieren	Köln	Kölner Stadt Anzeiger	15.01.2019	80,7	116,1	44%	Ursachen: Baupreissteig., Umplanung in der Bauphase, Fehlende Fachleute in der		https://www.ksta.de/koeln/innenstadt/koelner-bildungs-campus-kosten-fuer-schulbauten-am-klin...
17	Hansa Gymnasium	Köln	Kölner Stadt Anzeiger	16.01.2019			73%	Bauverwaltung, lange Zeitspanne zw. Projektbeschluss und Realisierung		https://www.ksta.de/koeln/innenstadt/koelner-bildungs-campus-kosten-fuer-schulbauten-am-klin...
18	Abendgymnasium	Köln	Kölner Stadt Anzeiger	17.01.2019			63%			https://www.ksta.de/koeln/innenstadt/koelner-bildungs-campus-kosten-fuer-schulbauten-am-klin...
19	Michelberg-Gymnasium kaputtsaniert: Wie ein verpfuschter Schulbau eine Stadt zu ruinieren droht	Geislingen	news4teachers.de	16.02.2020				Schaden 25-37 Mio. €, Leuchtturmprojekt und Ökobau nach Sanierung in 2016		https://www.news4teachers.de/2020/02/drohende-pleite-regierungspraesidium-verhindert-sanien...
19	Michelberg-Gymnasium Gy bei „Mario Barth deckt auf": „Eine Schule, die kaputtsaniert wurde"	Geislingen	RTL	09.04.2020				jetzt von Schließung bedroht (Einsturzgefahr), schwere Brandschutzprobleme		https://www.swp.de/suedwesten/staedte/geislingen/michelberg-gymnasium-geislingen-migy-bei-...
20	Kulturhaus Laubusch: Brandschutz wird deutlich teurer	Laubusch	Lausitz LR-online	08.05.2019	0,1	0,4	207%	Brandschutzauflagen		https://www.lr-online.de/lausitz/hoyerswerda/kulturhaus-laubusch-brandschutz-wird-deutlich-teu...
21	Schulerweiterung verzögert sich erneut	Radebeul	Sächsische SZ DE	27.11.2019				Weil die Kosten zu explodieren drohten, wird der Bau neu ausgeschrieben		https://www.saechsische.de/plus/schulerweiterung-verzoegert-sich-erneut-5144412.html
22	Schulneubau im Grimmaer Ortsteil Böhlen wird teurer	Grimma	Leipziger Volkzeitung	15.04.2019	8,4	10,5	25%	Klage u.a.		https://www.lvz.de/Region/Grimma/Schulneubau-im-Grimmaer-Ortsteil-Boehlen-wird-teurer
23	Nach Verzögerung und Kostenexplosion: Endlich gehen die Bauarbeiten in Schenefeld voran	Schenefeld	Hamburg Abendblatt	27.07.2019						https://www.abendblatt.de/region/pinneberg/article226500311/Endlich-gehen-die-Bauarbeiten-in...
24	Kostenexplosion bei öffentlichen Bauten	Hamburg	DIE WELT	13.08.2009				Von 217 Projekten werden 63 teurer, nur 18 wurden günstiger. Fast jedes dritte Baup...		https://www.welt.de/regionales/hamburg/article4316476/Kostenexplosion-bei-oeffentlichen-Baut...
25	- Grunderneuerung der Schule am Falkenberg	Hamburg	DIE WELT	13.08.2009	2,4	15,0	525%	Bauprojekt in Hamburg wird teurer als veranschlagt. Schulbau ist besonders		https://www.welt.de/regionales/hamburg/article4316476/Kostenexplosion-bei-oeffentlichen-Baut...
26	- Abriss und Neubau Grundschule Barlsheide	Hamburg	DIE WELT	13.08.2009	6,6	18,0	173%	oft betroffen. Auch die Kosten für Krankenhausbauten laufen aus dem Ruder		https://www.welt.de/regionales/hamburg/article4316476/Kostenexplosion-bei-oeffentlichen-Baut...
27	- Um- und Neubau der Gewerbeschule	Hamburg	DIE WELT	13.08.2009	1,9	7,8	311%			https://www.welt.de/regionales/hamburg/article4316476/Kostenexplosion-bei-oeffentlichen-Baut...
28	Berliner Schulbauoffensive: Kosten für Schulbau explodieren	Berlin	Berliner-Zeitung	04.11.2018	1.200,0	1.700,0	42%	Kostenbedarf steigt von 1,2 auf 1,7 Mrd. €;		https://www.berliner-zeitung.de/mensch-metropole/berliner-schulbauoffensive-kosten-fuer-schul...
29	- Neubau Schulzentrum Adlersdorf mit Grund- + Sekundarschule	Berlin	Berliner-Zeitung	04.11.2018	63,0	100,6	60%			https://www.berliner-zeitung.de/mensch-metropole/berliner-schulbauoffensive-kosten-fuer-schul...
30	- Umbau Sollingschule	Berlin - Marienfelde	Berliner-Zeitung	04.11.2018	13,0	45,0	246%			https://www.berliner-zeitung.de/mensch-metropole/berliner-schulbauoffensive-kosten-fuer-schul...
31	- Sanierung B.Traven-Schule	Berlin-Spandau	Berliner-Zeitung	04.11.2018	10,0	54,0	440%			https://www.berliner-zeitung.de/mensch-metropole/berliner-schulbauoffensive-kosten-fuer-schul...
32	Grundstein im Erdgeschoss	Offenbach	op-online	25.08.2011	250,0	334,0	34%	Kostenbedarf 32 Schulen steigt		https://www.op-online.de/offenbach/grundstein-erdgeschoss-1375177.htm
33	Millionen mehr für Gesamtschule Kleve werden akzeptiert	Kleve	NRZ	04.11.2019		+18		Torf im Untergrund und Starkregen		https://www.nrz.de/staedte/kleve-und-umland/millionen-mehr-fuer-gesamtschule-kleve-werden-a...
34	Kostenexplosion Gesamtschule Stadtmitte	Mönchengladbach	fwg.in-mg.de	07.09.2013			700%			https://www.fwg-in-mg.de/index.php/schule/114-kostenexplosion-gesamtschule-stadtmitte
35	Bis 2022 soll das Viereck stehen	Kell/Trier	Volksfreund	22.03.2018	8,8	12,5	42%	Dabei wird es nicht bleiben		https://www.volksfreund.de/region/trier-trierer-land/schule-in-kell-kostet-nun-12-5-millionen-eur...
36	Lippstädter Architekt nimmt Stellung zur Kostenexplosion bei Wickeder Sekundarschule	Lippstadt	wickede.ruhr	29.05.2015	5,3	6,3	19%	400T€ liegen im üblichen Rahmen; zusätzliche Nutzerwünsche		https://www.wickedepunktruhr.de/heimat-online/Aktuelle_Meldungen/2015-05-28_Sekundarsch...
37	Affären beschäftigen Berner Rathaus	Berne	NWZ	19.06.2008	2,0	3,0	50%	1. Kostenschätzung war politischer Preis		https://www.nwzonline.de/berne/affaeren-beschaeftigen-berner-rat-haus_a_3,1,27889342.html
38	Einblicke in das Triptis Geheimpapier - Kostenexpolision auf der Schulbaustelle	Triptis	Ostthüringer Zeitung	13.09.2017				1. Zwischenbericht zur Kostenexposion 13 S. wird nicht herausgegeben.		https://www.otz.de/politik/einblicke-in-das-triptis-geheimpapier-id223212181.html
39	Neubau Rossert-Grundschule	Kelkheim	Prof. Berner	17.11.2015	7,4	9,6	30%	Baukostenverschwendung im Schulbaubereich 1,76 Mrd.€		https://docplayer.org/61573534-Universitaet-stuttgart-kostensteigerungen-bei-oeffentlichen-bauv...
40	Umbau und Sanierung Schulzentrum Menden	St. Augustin	Prof. Berner	17.11.2015	4,1	7,3	78%	Gründe: Trennung Verantwortung Palnung + Umsetzung; Defizite in der Planung		https://docplayer.org/61573534-Universitaet-stuttgart-kostensteigerungen-bei-oeffentlichen-bauv...
41	Sanierung Schule Kroonhorst	Hamburg	Prof. Berner	17.11.2015	2,6	10,6	314%	und Bauausführung, mangelhafte Kostenschätzung/-management		https://docplayer.org/61573534-Universitaet-stuttgart-kostensteigerungen-bei-oeffentlichen-bauv...
42	Sanierung 3 Schulen	Magdeburg	Stadtelternrat	06.09.2019	29,3	35,8	22%	Preissteigerungen wegen langem Planungszeitraum über 4 Jahre		https://stadtelternrat-magdeburg.de/preisexplosion-im-schulbau/
43	Ein ganz normales Gymnasium - für 94 Millionen Euro	Kirchheim	Süddeutsche	01.06.2019	75,0	94,0	25%	1. Kostenobergrenze 75 Mio €; 1.Kostenschätzung 88 Mio.€		https://www.sueddeutsche.de/muenchen/landkreismuenchen/landkreis-muenchen-kirchheim-sch...
44	- Zitat Philip Leistner, Fraunhofer Institut. Kongress Zukunftsraum Schule							Schulbauten idR um 80% teurer als urspr. geplant; Ø Kosten: 3-4T€/qm Nutzfläche		https://www.sueddeutsche.de/muenchen/landkreismuenchen/landkreis-muenchen-kirchheim-sch...
45	Schulbau wird noch teurer: 39,5 Millionen Euro - Karlsfeld muss Kreditaufnahme erhöhen	Karlsfeld	Merkur	12.07.2020	20,0	39,5	98%	2015: unter 20 Mio. erste Kostenschätzung; Mitte 2019: 34 Mio. €		https://www.merkur.de/lokales/dachau/karlsfeld-ort28903/schule-krenmoosstrasse-kralsfeld-neub...
46	Lernort Horrem	Dormagen	NGZ Online	05.10.2018	9,8	12,7	30%	30 Prozent Risikozuschlag		https://rp-online.de/nrw/staedte/dormagen/dormagen-mehrkosten-und-verzoegerung-beim-umba...
47	Sekundarschule	Dormagen	NGZ Online	15.11.2019	8,2	15,5	90%			https://rp-online.de/nrw/staedte/dormagen/dormagen-mehrkosten-und-verzoegerung-beim-umba...
48	Kosten für Bochumer Gesamtschule explodieren	Bochum	lokalkompass.de	07.03.2020	21,0	50,0	138%	Neuer Kostenrahmen 40-50 Mio €		https://www.lokalkompass.de/bochum/c-politik/kosten-fuer-bochumer-gesamtschule-explodieren...
49	Erhebliche Kostensteigerung beim neuen Berufskolleg	Remscheid	waterboelles.de	11.02.2020	21,0	30,0	43%	Gestiegene Baukosten und weitere Präzisierungen der Planung		https://www.waterboelles.de/archives/25944-Erhebliche-Kostensteigerung-beim-neuen-Berufskoll...
50	Neubau der Grund- und Mittelschule Rott: Kosten marschieren nach oben	Rott am Inn	Wasserburg24.de	19.07.2019	16,5	18,4	12%			https://www.wasserburg24.de/wasserburg/region-wasserburg/rott-am-inn-ort60205/rott-erneut-kr...
	Durchschnittliche Erhöhung der Baukosten						116%			

ANNEX D

European Comparative PPP Study - German projects - Definitions				
A	DIN 18960	Usage costs	Total of cost groups CG 100-400	Maintenance costs
1	CG 100	Capital costs	Amortization and interest costs	
2	CG 200	Property management costs	Personal costs in context with the facilities management including material costs and overhead costs	Partial property management costs
3	CG 300	Operating costs	Total of cost groups CG 310-390	
4	CG 310	Energy management costs	Electricity, heating/cooling, water	
5	CG 320	Waste management costs	Waste water, rainwater, waste	
6	CG 330	Cleaning of the building	Regular and basic cleaning of the inuilding, including windows	
7	CG 340	Cleaning and care of the outside facilities	Cleaning of the outdoor facilities including care of the green area	
8	CG 350	Inspection&maintenance	Part of the maintenance costs	Inspection & maintenance
9	CG 360	Security and controlling services	Property and personal protection, control tasks due to public law	
10	CG 370	Duties and contributions	E.g. assurance costs, debt guarantees	
11	CG 390	Other operating costs	E.g. catering, mensa/cafeteria, legal and auditing costs	
12	CG 400	Repair costs	Renovation and replacement of outworn components, including improvements, aesthetic repairs, part of the maintenance costs	Repair costs
B		**Revenues and residual value**		
13		Operating revenues	E.g. parking fee of an underground parking	
14		Residual value	Expected residual value of the asset at the end of the life cycle period minus not amortized residual debts	
C		**Life cycle costs**	**Usage costs minus revenues and residual value**	

Bibliography

Bahr, Carolin, Realdatenanalyse zum Instandhaltungsaufwand öffentlicher Hochbauten", Dissertation 2008, University of Karlsruhe

Bahr Carolin / Lennerts, Kunibert, Lebens- und Nutzungsdauern von Bauteilen, Gutachten im Rahmen des Forschungsprogramms Zukunft Bau, February 2010

Bewer, Rebecca / Iding, Andreas / Kronsbein, Dirk, Richtig handhaben: das Rücklagenkonto bei ÖPP-Projekten, PPP Jahrbuch 2012, p. 163 ff.

BKI Baukosten 2015 Neubau, part 1, Statistische Kostenkennwerte für Gebäude, Stuttgart, 2015

BKI Baukosten 2014 Neubau, part 1, Statistische Kostenkennwerte für Gebäude, Stuttgart, 2014

BKI Objektdaten, NKS Nutzungskosten, Kosten von Bestandsimmobilien und statistische Kostenkennwerte, Stuttgart, 2015

Bogenstätter, Ulrich, Property Management und Facility Management, Munich, 2008

Brand, Stephan / Steinbrecher, Johannes, Kommunaler Investitionsrückstand bei Schulgebäuden erschwert Bildungserfolge, KfW Research no. 143, 24 September 2016

Bürsch, Michael, Eine neue Arbeitsteilung, in: Pauly, Lothar (ed.), PPP - Das neue Miteinander, Hamburg, 2006, p. 163 ff.

Bürsch, Michael / Funken, Klaus, Kommentar zum ÖPP-Beschleunigungsgesetz, Frankfurt am Main, 2007

Bundesgutachten PPP im Öffentlichen Hochbau, Berlin, August 2003, Gutachtergruppe Price Waterhouse Coopers, Freshfields Bruckhaus Deringer, VBD, Bauhaus University Weimar, Creative Concept; vol. I: Leitfaden PPP im Öffentlichen Hochbau, vol. II: Rechtliche Rahmenbedingungen, vol. III: Wirtschaftlichkeitsuntersuchung, vol. IV: Sammlung und systematische Auswertung zu Informationen zu PPP-Beispielen, vol. V: Strategie/Taskforces

Bundesministerium der Finanzen, Das System der öffentlichen Haushalte, 2015

Bundesministerium des Innern, für Bau und Heimat, Bewertungssystem Nachhaltiges Bauen (BNB), Unterrichtsgebäude (BNB_UN 2.1.1), Gebäudebezogene Kosten im Lebenszyklus, edition: 12 December 2017

Bundesministerium für Verkehr, Bau und Stadtentwicklung, Leitfaden Wirtschaftlichkeitsuntersuchungen (WU) bei der Vorbereitung von Hochbaumaßnahmen des Bundes., 2nd ed. 2013

Bundesrechnungshof, Bericht an den Haushaltsausschuss des Deutsches Bundestages nach § 88 Abs. 2 BHO über Öffentlich Private Partnerschaften (ÖPP) als Beschaffungsvariante im Bundesfernstraßenbau, Bonn, 4 June 2014 (ref.: V3-2013-5166)

Bundesrechnungshof, Bericht nach § 88 Absatz 2 BHO, Erfolgskontrollen als Voraussetzung für eine wirkungsorientierte Haushaltsführung, 26.07.2023 (Gz: V5 – 0001610)

Christen, Jörg, in: PPP-Handbuch, Leitfaden für Öffentlich-Private Partnerschaften, ed.: Bundesministerium für Verkehr, Bau und Stadtentwicklung sowie Deutscher Sparkassen- und Giroverband, 2nd ed. 2009, Chapter 1, Einführung, p. 7 ff.

Christen, Jörg, Sind PPP-Projekte wirtschaftlicher als Eigenbau? Praxiserfahrungen zum Wirtschaftlichkeitsnachweis bei PPP-Projekten, in: Gralla, Mike/ Sundermaier, Matthias (eds.) Innovationen im Baubetrieb, Commemorative publication for Prof. Dr-Ing. Udo Blecken on his 70th birthday, Cologne, 2011, p. 599 ff.

Christen, Jörg / Dusch, Sandra, Wirtschaftliche Dotierung von Instandhaltungsbudgets am Beispiel von 370 konventionellen und 7 PPP-Kitas, in: Kessel, Tanja/Gawlitta, Marcel/Hilbig, Corinna/Walther, Martina (eds.), Commemorative publication for Prof. Dr Dieter Jacob, Frankfurt, 2015, p. 385

Christen, Jörg / Guerriero, Raffaele: Wirtschaftlichkeit von Bau und Instandhaltung bei 880 konventionellen Schulen und 50 PPP-Schulprojekten; Mainz, 2019 [PPP School Study (2019)]

Christen, Jörg / Rüdiger, Danny, Wirtschaftlichkeitsuntersuchung zum PPP-Berufskolleg Duisburg, Mainz, 2024

Christen, Jörg / Strobel, Michael / Thomas, Ise, Südbad Trier, Vom Denkmal zum PPP-Pilotprojekt, Mainz, 2010

Clement, Wolfgang, Ärmel hoch und weiter, Verbesserung der PPP-Rahmenbedingungen - eine Daueraufgabe, in: Pauly, Lothar (ed.), PPP - Das neue Miteinander, Hamburg, 2006, p.157 ff.

Denker, Philipp / Kunzmann, Melanie / Vogt, Henrik, in: ÖPP Deutschland AG (ed.), Wirtschaftlichkeitsuntersuchungen für Öffentlich-Private Partnerschaften im Hochbau, Analyse und Potenziale, ÖPP-Schriftenreihe vol. 18, Berlin, 2016

Deutsches Institut für Urbanistik (DIFU), Public Private Partnership Projekte - eine aktuelle Bestandsaufnahme in Bund, Ländern und Kommunen, Berlin, 2005

Dusch, Sandra, Wirtschaftliche Dotierung von Instandhaltungsbudgets, Masterarbeit im Studiengang Immobilienprojektmanagement, Fachhochschule Mainz, November 2013

EUROSTAT, PPP – A Guide to the Statistical Treatment of PPPs, Luxembourg, September 2016

Funken, Klaus, in SPD-Bundestagsfraktion (ed.), Öffentlich Private Partnerschaften – eine Zwischenbilanz im Jahre 2009, document no. 09/09, SPD-Bundestagsfraktion, Berlin, June 2009, p. 7 ff.

FMK Guidelines, Leitfaden Wirtschaftlichkeitsuntersuchungen bei PPP-Projekten, 2006, Homepage Bundesministerium für Umwelt, Naturschutz, Bau und Reaktorsicherheit, search term: "Bundeseinheitlicher Leitfaden für PPP-Wirtschaftlichkeitsuntersuchungen"

Gemeinsamer Erfahrungsbericht der Rechnungshöfe des Bundes und der Länder zur Wirtschaftlichkeit von ÖPP-Projekten, Wiesbaden, 2011

Gornig, Martin / Michelsen, Claus, Kommunale Investitionsschwäche: Engpässe bei Planungs- und Baukapazitäten bremsen Städte und Gemeinden aus, DIW Wochenbericht no. 11/2017, p. 211 ff.

Gottschling, Ines / Pickenäcker, Birgit Anne, PPP evaluieren, in: Evaluation von PPP-Schulprojekten, PPP-Newsletter der PPP Task Force im BMVBS, special edition 2008

Grabow, Busso / Hollbach-Grömig, Beate / Schneider, Stefan, PPP und Mittelstand, Untersuchung von 30 ausgewählten PPP-Hochbauprojekten in Deutschland, 2008, DIFU-Studie im Auftrag der PPP Task Force im BMVBS und der PPP Task Force NRW

Großmann, Achim, PPP-Initiative der Bundesregierung, press release, 7 September 2005, www.baulinks.de

Hopfe, Jörg / Napp, Hans-Georg/Bergmann, Sebastian / Keckeis, Lothar, in: PPP-Handbuch, Leitfaden für Öffentlich-Private Partnerschaften, ed.: Bundesministerium für Verkehr, Bau und Stadtentwicklung sowie Deutscher Sparkassen- und Giroverband, 2nd ed. 2009, Chapter 4 Finanzierung, p. 164 ff.

Hoppenberg, Michael / Dinkhoff, Marc / Schäller, Sebastian, in: PPP-Handbuch, Leitfaden für Öffentlich-Private Partnerschaften, ed.: Bundesministerium für Verkehr, Bau und Stadtentwicklung sowie Deutscher Sparkassen- und Giroverband, 2nd ed. 2009, Chapter 3, Contract drafting, p. 65 ff.

Icha, Petra, Lauf, Thomas, Kuhs, Gunter, Bundesumweltamt (ed.), Entwicklung der spezifischen Treibhausgas-Emissionen des deutschen Strommix in den Jahren 1990 – 2021, Dessau-Roßlau, April 2022

Institut für Demoskopie Allensbach, Die Zufriedenheit mit ÖPP-Projekten im Schulbereich aus Sicht von Auftraggebern, Schulleitern und Elternvertretern, 2010

Jacob, Dieter, Wirtschaftlichkeit von Public Private Partnership am Beispiel Schulen, Freiberg, 2005

Kalusche, Wolfdietrich, Frühzeitige Ermittlung der Baunebenkosten bei der Gebäudeplanung, in: BKI Baukosten Gebäude, Statistische Kostenkennwerte 2014, p. 44

Klingenberger, Jens: Ein Beitrag zur systematischen Instandhaltung von Gebäuden, Dissertation, Darmstadt, 2007

Kessel, Tanja / Schottel, Kristin / Peuker, Swaantje, ÖPP-Praxistest 2.0: Evaluierung der Nutzungsphase, Studie im Auftrag des Hauptverbands der Deutschen Bauindustrie e.V., Berlin, October 2016

KfW-Kommunalpanel 2024, in Kooperation mit dem Deutschen Institut für Urbanistik, Frankfurt/Main, May 2024

KGSt, Kommunale Gemeinschaftsstelle für Verwaltungsmanagement (KGSt): Hochbauunterhaltung, Richtwerte und Gestaltungsvorschläge zur Mittelbemessung, Maßnahmenplanung und Mittelbereitstellung, Report 9, Cologne, 1984

Knop, Detlef (ed.), Public Private Partnership Jahrbuch 2013, Frankfurt/Main, 2013

Koalitionsvertrag 2021 – 2025 zwischen der Sozialdemokratischen Partei Deutschlands (SPD), BÜNDNIS 90 / DIE GRÜNEN und den Freien Demokraten (FDP)

Kommunale Gemeinschaftsstelle für Verwaltungsmanagement (KGSt): Instandhaltung Kommunaler Gebäude, report 7, Cologne, 2009

Krajewski, Gunther, PPP – Heilslehre oder Grundsatzdebatte? Ordnungspolitische Instrumente werden durch PPP in Frage gestellt, DAB 2002, p. 1-

Krawczyk, Stanislaus, Erfahrungsbericht aus Eppelheim, in: Evaluation von PPP-Schulprojekten, PPP Newsletter der PPP Task Force im BMVBS, special edition 2008, p. 2-4

Landesrechnungshof Hessen - Der Präsident des Hessischen Rechnungshofs – Überörtliche Prüfung kommunaler Körperschaften, 182nd PPP audit review Offenbach district, final report dated 26 March 2015

Landesrechnungshof Hessen, 18th report, PPP-Projekt Schulen des Landkreises Offenbach, summary report dated 6 February 2009, 1. 151 ff.

Landesrechnungshof Rheinland-Pfalz, Prüfbericht zum PPP-Projekt Südbad Trier, 24 January 2011, p. 44 ff.

Landesrechnungshof Sachsen-Anhalt, Jahresbericht 2013, part 2, p. 66 ff. no. 4: Umsetzung von Schulprojekten im Vergleich zwischen ÖPP und konventioneller Beschaffungsvariante

Lennerts, Kunibert; Pfründer, Uwe; Bahr, Carolin.: Lebenszyklusorientierte ganzheitliche Unterhalts- und Instandhaltungsstrategien für Schulen. Leaflet "Kostenoptimierung bei Schulgebäuden", EnergieEffizienzAgentur.E2A Rhein-Neckar, Ludwigshafen, December 2006

Littwin, Frank, PPP - kein Allheilmittel, aber mehr Kostentransparenz und verstärktes wirtschaftliches Handeln, in: Ifo Schnelldienst 24/2006, p. 5

McCleary, Boyd, PFI in Großbritannien – Erfahrungen mit der privaten Finanzierung öffentlicher Projekte, DAB 2002, p. 16

Partnerschaft Deutschland, ÖPP-Mustervertrag für ein Inhabermodell, ÖPP-Schriftenreihe, vol. 15

Pols, Helge, Rechtliche Grundlagen und Rahmenbedingungen für PPP, Augsburg, 2006

PPP Task Force des Bundes im Bundesministerium für Verkehr, Bau und Stadtentwicklung, PPP-Schulstudie, Leitfaden 1: Chancen und Risiken von PPP in den Neuen Bundesländern, Leitfaden 2: Kriterienkatalog PPP-Eignungstest für Schulen, Leitfaden 3: Outputorientierte Ausschreibungsunterlagen, Leitfaden 4: PPP-Wirtschaftlichkeitsuntersuchung, Leitfaden 5, Mustervertrag PPP-Inhabermodell / PPP-Mietmodell, Berlin, June 2007

PPP Task Force im Finanzministerium Nordrhein-Westfalen, Bericht zur Untersuchung der Auswirkungen von unterschiedlich umfangreichen Instandhaltungs- und Sanierungsmaßnahmen an kommunalen Gebäuden, Düsseldorf, 2011

PPP Task Force im Finanzministerium Nordrhein-Westfalen, Public Private Partnership im Hochbau, Vertragsrechtliche Aspekte am Beispiel von PPP-Schulprojekten, Düsseldorf, November 2005

Proll, R. Uwe / Drey, Franz, Die 20 Besten: PPP-Beispiele aus Deutschland, Cologne, 2006

Quaschning, Volker, Regenerative Energiesysteme, 11th ed., Munich, 2021

Rein, Stefan und Gottschling, Ines, Wirtschaftlichkeitsuntersuchungen im öffentlichen Hochbau, in: Kessel, Tanja / Gawlitta, Marcel / Hilbig,Corinna / Walther, Martina (ed.) Commemorative publication for Prof. Dr Dieter Jacob, Frankfurt, 2015, p. 277

Richtlinien für die Durchführung von Bauaufgaben des Landes Rheinland-Pfalz Ausgabe 2006, July 2014 edition

Ritter, Frank, Lebensdauer von Bauteilen und Bauelementen, Dissertation, Darmstadt, 2011

Ross, Frank-Wilhelm / Brachmann, Rolf / Holzner, Peter, Ermittlung des Bauwertes von Gebäuden und des Verkehrswertes von Grundstücken, Hannover, 1997, pp. 267-268

Scheel-Siebenborn, Axel, Ist Erfolg öffentlicher Hochbauprojekte objektiv messbar ?, in: Kessel, Tanja / Gawlitta, Marcel / Hilbig, Corinna / Walther, Martina (ed.), Commemorative publication for Prof. Dr Dieter Jacob, Frankfurt, 2015, p. 291

Scheidt, Konstantin, Betreiberhaftung bei der konventionellen und PPP-Instandhaltung, 2020, Münster/Mainz

Schmitz, Heinz / Krings, Edgar / Dahlhaus, Ulrich J. / Meisel, Ulli, Baukosten 2012/2013: Altbau

Senatsverwaltung für Bildung, Jugend und Wissenschaft, Die Senatorin, Handlungsrahmen Berliner Schulbau 2026, 8 June 2016

Schönfelder, Uwe Thomas: Verfahren zur Ermittlung des Abnutzungsvorrats von Baustoffen als Grundlage für Instandhaltungsstrategien am Beispiel der Gebäudehülle, Dissertation, Dortmund, 2010, p. 43

SPD-Bundestagsfraktion (ed.), Positionspapier der Projektarbeitsgruppe PPP, documents 04/2002

SPD-Bundestagsfraktion (ed.), Öffentlich Private Partnerschaften – ein Wegweiser für Kommunen, documents no. 01/04

SPD-Bundestagsfraktion (ed.), Das ÖPP-Beschleunigungsgesetz - ein Projekt der SPD-Bundestagsfraktion, documents no. 03/05

SPD-Bundestagsfraktion (ed.), Öffentlich Private Partnerschaften - eine Zwischenbilanz im Jahre 2009, documents no. 09/09

Stächele, Willi, Öffentlich-Private Partnerschaften (ÖPP) im staatlichen Hochbau, Finanzministerium in Baden-Württemberg, September 2010

Steinbrück, Peer, Wir brauchen eine PPP-Kultur, in Pauly, Lothar (ed.): Das neue Miteinander. Public Private Partnership für Deutschland, Hamburg, 2006, p. 145 ff.

Stiepelmann, Heiko, PPP - Der partnerschaftliche Weg aus dem öffentlichen Investitionsstau, in: Knop (ed.): Public Private Partnership. Jahrbuch, Wiesbaden, 2004, p. 73 ff.

Stolpe, Manfred, Investitionen mobilisieren, staatliches Handeln modernisieren, in: Pauly, Lothar (ed.), PPP - Das neue Miteinander, Hamburg, 2006, p. 65 ff.

Tiefensee, Wolfgang, Anteil von PPP an öffentlichen Investitionen soll weiter steigen, Erster Erfahrungsbericht zu Public Private Partnership, 17 April 2007, Short link: www.bauingenieur24.de/url/700/1821

Tomm, Arwed / Rentenmeister, Oswald / Finke, Heinz Geplante Instandhaltung, Landesinstitut NRW für Bauwesen und angewandte Bauschadensforschung, Aachen, 1995

Truger, Achim, Die Goldene Regel für öffentliche Investitionen als Ausweg aus der Wirtschaftskrise im Euroraum, in: Junkernheinrich, Martin/Korioth, Stefan / Lenk, Thomas / Scheller, Henrick / Woisin, Matthias (ed.), Jahrbuch für öffentliche Finanzen 2-2016, p. 347 ff.

VDI-Richtlinie 2067 page 1 "Wirtschaftlichkeit gebäudetechnischer Anlagen - Grundlagen und Kostenberechnung", edition: September 2010

Verweij, Stefan / van Meerkerk, Ingmar / Casady, Carter B., Assessing the Performance Advantage of Public-Private Partnerships, Edward Elgar Publishing, 2022

Walter, Peter, PPP für Schulen im Kreis Offenbach, in: Knop, Detlef (ed.): Public Private Partnership Jahrbuch, Wiesbaden, 2004, p. 84 ff.

Weber, Martin / Schäfer, Michael / Hausmann, Friedrich Ludwig, Praxishandbuch Public Private Partnership, 2nd edition, Munich, 2018

Wissenschaftlicher Beirat beim Bundesministerium der Finanzen, Chancen und Risiken Öffentlich-Privater Partnerschaften, Gutachten, Berlin, February 2016

Zehle, Sonja, Schulprojekte evaluieren! – PPP Newsletter 01/2008 der PPP Task Force im Bundesministerium für Verkehr, Bau und Stadtentwicklung, p. 3

Notes

[1] See Verweij/van Meerkerk/Casady, Assessing the Performance Advantage of Public Private Partnerships, p. 217 ff.

[2] See Christen and Jörg. PPP-Handbuch, p. 15 ff.; PPP-Bundesgutachten 2003, vol. IV, p. 8 f with the "initial impression" from positive project reports from 46 PPP forerunner projects in the 1990s, in which savings of 20% over the status quo were reported as a result of linking responsibility for construction and financing. In a DIFU survey in 2005, 83% of the 231 local authorities and 80% of the 63 districts surveyed stated that the main reason for implementing PPP projects was the expectation of increased efficiency.

[3] See report by the German Federal Audit Office dated 4 June 2014 on the inefficiency of PPP motorways in the amount of €1.9 billion, which arose from 17% higher private financing costs (p. 15); these additional costs could not be offset by productivity gains in construction costs due to inflexible technical standards (p. 17). An assessment of qualitative differences between PPP project financing and conventional financing was not carried out, nor was an assessment of the risk of late delivery and deadline overruns (cf. report of the European Court of Auditors of 3 September 2013, where the late delivery rate for 6 German motorway projects was 26% and the deadline overruns 59%, cf. note below nr. 23)

[4] See Gemeinsamer Erfahrungsbericht der Rechnungshöfe des Bundes und der Länder zur Wirtschaftlichkeit von ÖPP-Projekten, 2011.

[5] See *Die Zeit* (25 October 2012): "Well calculated".

[6] See *Welt am Sonntag* (15 February 2015): "Philosopher's stone?": Quote from Prof. Dr Lars Feld (German Council of Economic Experts): "There is a risk of generating debt that is not in the budget" and Dr Anton Hofreiter (MP) on the Fratzscher Commission's proposals: "This is nothing more than a way to subsidise life insurers by bypassing the debt brake with overpriced interest rates." This criticism does not generally apply to local authorities and therefore to the majority of German PPP projects to date, provided that the municipal budget is managed according to the double-entry method.

[7] An analysis of the German government's PPP project database (www.ppp-projektdatenbank.de), the Federation of the German Construction Industry's PPP platform (www.oepp-plattform.de) and other sources (VIFG — Verkehrsinfrastrukturfinanzierungsgesellschaft mbH's website, BWI-Bau PPP newsletter, press releases, etc.) reveals that as of December 2023 there were 281 projects with an investment volume of €9.6 billion in the building construction sector and 26 projects with an investment volume of €7.8 billion in the transport sector. That is a total of 307 projects with construction investment costs of € 17.4 billion. This does not include projects for which only minor operating services are included in the contract in addition to planning, construction and financing. The education sector is the largest sub-sector to date with 110 projects.

[8] See Gemeinsamer Erfahrungsbericht der Rechnungshöfe des Bundes und der Länder zur Wirtschaftlichkeit von ÖPP-Projekten, September 2011; one of the 18 projects analysed is a P2P (public-public partnership) school project; Saxony-Anhalt State Audit Office, Annual Report 2013, part 2, p. 66 ff., no. 4: Umsetzung von Schulprojekten im Vergleich zwischen ÖPP und konventioneller Beschaffungsvariante.

[9] The applicable period for determining the respective price increase was 2007 to 2021.

[10] See BKI-Objektdaten — Sonderband Schulen, Stuttgart 2017, p. 294 ff.

[11] The price risk is generally the responsibility of the contracting authority.

[12] See Quaschning and Volker (2021). Regenerative Energiesysteme, 11th edition, Munich; https://www.volker-quaschning.de/datserv/CO2-spez/index.php.

[13] See Icha, Petra, Lauf, Thomas, Kuhs and Gunter, Bundesumweltamt (ed.) (2022). Entwicklung der spezifischen Treibhausgas-Emissionen des deutschen Strommix in den Jahren 1990 – 2021. Dessau-Roßlau, p. 11.

[14] See Scheidt and Konstantin (2020). Betreiberhaftung bei der konventionellen und PPP-Instandhaltung. Bachelor's thesis, FH Münster/HS Mainz, p. 30 ff.

[15] See Scheidt, Konstantin, Betreiberhaftung, loc. cit.

[16] See KfW-Kommunalpanel 2024, p. 14.

[17] See Gornig, Martin / Michelsen, Claus, Kommunale Investitionsschwäche: Engpässe bei Planungs- und Baukapazitäten bremsen Städte und Gemeinden aus, DIW Wochenbericht Nr. 11.2017, S. 211 ff.; Frankfurter Rundschau vom 16.3.2017: Milliarden unverplant. The study by the German Institute for Economic Research (DIW) from 15 March 2017 on the weakness of municipal investment points to considerable bottlenecks in municipal planning and construction capacities: According to the report, the number of employees dealing with construction issues in the municipal administrations of cities and municipalities fell by around 45 per cent between 1991 and 2015. Although the federal government set up a municipal investment promotion fund for financially weak

municipalities totalling EUR 3.5 billion in 2015, only EUR 145 million of this has been called up and EUR 1.8 billion has been planned.

[18] See the practical example in Annex A, pp. 69, 71 and 74 for the calculation of life cycle costs excluding and including risk costs.

[19] See paragraph 158.

[20] See paragraph 162 f.

[21] See KfW-Kommunalpanel 2024, p. 14.

[22] Here, the construction costs of all 18 PPP projects are totalled and compared with the sum of all comparable conventional BKI construction costs. The timing of the costs incurred (PPP construction costs from 2004, 2010, 2018, etc.) being different has not yet been considered.

[23] For information on the certainty of deadlines for PPP motorway projects, see the German Federal Audit Office's report of 4 June 2014. Although adhering to construction schedules or running ahead of them is attested here, shorter construction times could also be achieved using conventional methods if the administration were staffed accordingly (p. 27). No statements were made on cost certainty. In a report on conventional trunk road construction dated 14 April 2014, the German Federal Audit Office found that many federal trunk road projects were more expensive than planned. This was followed by an analysis of the causes (the Federal Ministry of Transport and Digital Infrastructure did not allocate budget funds in line with requirements, excessive planning, inadequate project preparation, absence of soil surveys, incorrect specifications) and 28 pages of recommendations for improving conventional cost management. No information is provided on the amount of the identified cost increases. In its audit report on the efficient use of EU funding for 24 European trunk road projects dated 3 September 2013, the European Court of Auditors found that in the six German projects it examined there were subsequent cost increases of 26% (EU average: 21%) over the contractually agreed construction costs, and the agreed deadlines were exceeded by 59% (EU: 41%).

[24] A survey conducted by the PPP Competence Centre Rhineland-Palatinate of the 35 subcontractors involved in the Südbad Trier PPP pilot project in the summer of 2010 revealed that 83% of the PPP companies' construction contracts were awarded to subcontractors in the surrounding region. The survey paints a positive picture overall: For example, 41% rated the progress of the project as "good," 22% as "very good" and the same number as "satisfactory." The findings are very similar on the topics of deadlines, remuneration and payment practices; see Christen, Strobel and Thomas (2010). Südbad Trier — Vom Denkmal zum Pilotprojekt, p. 23 f.

[25] The agreed schedule for Südbad Trier was honoured despite two severe winters and a number of challenges in the construction process (such as water ingress in the technical room and unexpected ceiling damage in the entrance building). Comparable issues in the conventional construction process could easily have caused construction to be halted for several months, additional costs incurred by a new tender and the loss of revenue from an entire swimming season (see Christen, Strobel and Thomas (2010). Südbad Trier — Vom Denkmal zum Pilotprojekt, p. 18).

[26] This means that for the 16 PPP projects analysed here, the maintenance budget is significantly higher than that of the 34 new-build projects in the PPP school study (1.2% per annum).

[27] The difference between PPP and the BKI model is minimal, although at 1.7% per annum the BKI maintenance budget as a percentage of restoration costs is higher than the average PPP budget (1.6% per annum). One reason for this is the special circumstances of two BKI projects and the other is the assumption that the rate at which the useful life is extended as maintenance budgets increase is somewhat slower than the rate at which it is shortened as the budget decreases.

[28] The significant increase in the percentage difference compared to the result for usage costs is because the assessment basis for the comparison (usage costs minus residual value) is significantly smaller.

[29] See Landtag Rheinland-Pfalz, LT-Drs. 17-6641 of 17 July 2018, Kleine Anfrage zu PPP mit dem Unternehmen Arvato: "The state government does not believe that it is possible to make a general statement that PPP projects are fundamentally more economical than the conventional implementation of projects. The rationale behind PPP projects is the economic implementation of the life cycle approach. The objective must be to optimise the investment and subsequent costs of public infrastructure projects in a sustainable manner by linking the organisational planning, construction, operation, financing and, if necessary, use. Using public-private partnerships is only warranted in individual cases if the economic viability of costs, qualities and processes is guaranteed over the intended usage period and this is also confirmed during operation."

[30] In the case of the first-generation property leases, the investment costs were fully amortised via the lease instalments during the term of the lease. In order to improve the liquidity burden during the term of the contract, the second-generation property leases were then switched to partial amortisation.

[31] See PPP-Schulstudie (2019), paragraph 180 ff.

[32] See KfW-Kommunalpanel 2024, p. 14.

[33] According to the results of this study, PPP construction costs account for 37% of PPP usage costs over 25 years. With building construction costs of € 9.6 billion, this results in estimated utilisation costs of € 26 billion. Assuming an equal percentage share of construction costs and utilisation costs, the estimated utilisation costs over 25 years for the 307 German PPP projects in building construction and transport amount to €46 billion.

[34] See for example BMF (2015). Das System der öffentlichen Haushalte, under G. Ergebnisorientierung des Haushalts, p. 84: "All measures should be subject to a performance review to verify the achieved results." In its report of 26 July 2023 (Performance reviews as a prerequisite for impact-oriented budget management), the German Federal Audit Office found that, contrary to the requirements of Section 7 of the Federal Budget Code, the federal authorities often carried out no or only inadequate performance reviews and that the extensive omission of performance reviews constitutes a serious breach of budgetary law.

[35] See Koalitionsvertrag der Bundesregierung zur 20. Legislaturperiode, line 5479 ff.: "When it comes to core functions of the state, implementation and financing by the state remain fundamental. Certain individual projects and procurements can be carried out within the framework of public-private partnerships. It has to be shown — taking into account the risks — by means of an economic viability study based on standardised criteria that it is more economical to implement a specific PPP project. Financial controlling and executive, parliamentary and public scrutiny must be ensured. The respective results, including the economic viability studies and awarded contracts, must be published in a transparent manner on the internet. The methodology for analysing the economic viability of PPP projects will be refined in line with existing recommendations of the German Federal Audit Office and adapted to the state of the art."